AUSTRALIAN
BUSTARD

AUSTRALIAN
BUSTARD

MARK ZIEMBICKI

CSIRO

PUBLISHING

National Library of Australia Cataloguing-in-Publication entry

Ziembicki, Mark.

Australian bustard/Mark Ziembicki.

9780643096110 (pbk.)

Australian natural history series.

Includes index.
Bibliography.

Bustards – Australia.

598.320994

Published by
CSIRO PUBLISHING
36 Gardiner Road, Clayton VIC 3168
Private Bag 10, Clayton South VIC 3169
Australia

Telephone: [+613] 9545 8555
Local call: 1300 788 000 (Australia only)
Fax: +61 3 9662 7555
Email: csiropublishing@csiro.au
Web site: www.publishing.csiro.au

Cover photos by Bruce Doran

Set in 10.5/14 Adobe Palatino, Optima and Stone Sans
Edited by Janet Walker
Cover and text design by James Kelly
Typeset by Desktop Concepts Pty Ltd, Melbourne
Printed by Ingram Lightning Source

CSIRO PUBLISHING publishes and distributes scientific, technical and health science books and journals from Australia to a worldwide audience and conducts these activities autonomously from the research activities of the Commonwealth Scientific and Industrial Research Organisation (CSIRO).

Jan26_RP_ILS

CONTENTS

PREFACE

Sighting an Australian bustard for the first time in the wild is a memorable experience for many people. I still vividly recall my own first encounter with these impressive birds. It was in the Lakefield National Park region of northern Queensland. As a southerner I vaguely knew of the bustard, but the large male standing his ground before us as we pulled up in our Toyota struck me by his size and stately appearance. The encounter was brief, but it left an indelible impression and that feeling of excitement one gets when sighting a new species, particularly one of such grandeur and mystery, for the first time. I wasn't to know it at the time, but our paths were to cross again in a decidedly more profound manner.

A few years later the bustard was to become the focus of my PhD research. In choosing a study subject, my intention was to focus on a single species that could be used as a model for examining the complex movement patterns made by many of Australia's birds. The bustard is considered a highly mobile species that undertakes widespread and dispersive movements in relation to highly variable environmental conditions. It also employs an interesting and highly specialised mating system, while its cultural significance to Aboriginal people across the country adds an additional element to its appeal. For such reasons the bustard, if you'll pardon the pun, seemed to fit the bill nicely as a potential study subject. To my surprise, it soon became obvious that very little was known about the species. The Australian bustard had never been formally studied in the field. Much of what was known was based on descriptive or anecdotal accounts by amateur naturalists and casual observers. Max Downes had undertaken some pioneering surveys in parts of the Northern Territory and South Australia, while the only detailed study of the bustard's behaviour was based on a captive population in southern Australia by Kate Fitzherbert in 1978. Much of the information making up this book is based on these pioneering studies and in large part on my own research of the species ecology in the tropical savannas of northern Australia. In the outback, most people know the bustard, where it is more commonly known as the bush or plains turkey. However, the species remains poorly known to most Australians. It is my intention in writing this book to shed some light on the species and convey some of the charm and mystery of this cryptic and lordly icon of the outback.

ACKNOWLEDGEMENTS

CSIRO Publishing provides an important avenue for publishing natural history and science publications that highlight Australia's unique and wonderful natural heritage. I would like to thank them for taking on the book and the opportunity to document here aspects of the cultural significance and biology of the Australian bustard in an easily accessible and quality format. I would especially like to thank John Manger, Tracey Millen and Janet Walker for their assistance in preparing this publication.

Although until recently the Australian bustard had not been formally studied in detail in the field, the pioneering work of Max Downes and Kate Fitzherbert, and the many observations made by numerous observers that are well summarised in the *Handbook of Australian, New Zealand and Antarctic Birds* (HANZAB) series, have significantly added to our knowledge of the species. In addition, several other closely related bustard species have been studied in detail by various researchers. I am indebted to their contributions and insights, many of which I draw on here. Much of the information presented in this book is based on research conducted for my PhD thesis while based in Darwin at the Biodiversity Conservation Unit (BCU) of the Northern Territory (NT) Department of Natural Resources, Environment and the Arts, and while enrolled at the School of Earth and Environmental Sciences at the University of Adelaide. These studies and this book were made possible by the substantial in-kind and logistical support of these institutions. I would like to thank all the staff at the Biodiversity Conservation Unit in Darwin, with particular thanks to Irene Rainey, Charmaine Tynan, Brooke Rankmore, Owen Price, Craig Hempel, Cameron Yates, Riikka Hokkinnen, Felicity Watt, Alaric Fisher, Brydie Hill and Jenni Low Choy for support and specific help at various times. I am particularly grateful for the support, encouragement and patience of John Woinarski (BCU) and David Paton (University of Adelaide). More than just academic supervisors, both are friends, role models and significant inspirations to wildlife biologists across the country and further afield.

Field studies for the PhD research were primarily conducted at the Douglas-Daly Research Farm and the Victoria River Research Station (Kidman Springs) in the Top End of the Northern Territory. Both stations are run by the NT Department of Regional Development, Primary Industry, Fisheries and Resources. Thanks to Peter O'Brien, Don Cherry, Tony Moran and all staff at these facilities. I am also grateful to the many volunteers who contributed their time in often trying circumstances in the pursuit of knowledge concerning bustards. Special thanks to Colin Bailey, Tony Dingwell, Bill Gordon, Claudia Franco, Luis Verissimo, Pedro Rocha and Jarrah. Financial support for field work was provided by a number of organisations. I am grateful to the Herman Slade Foundation as the principal supporter, and to all other contributors including Land & Water Australia (through the Science and Innovation Awards for Young People in Agriculture, Fisheries and Forestry), Stuart Leslie Bird Research Award, Australian Bird Environment Fund, the Wildlife Conservation Fund (SA Department of Environment and Heritage) and Bruce Doran (Australian National University). Financial and in-kind support was also provided by the Tropical Savannas Cooperative Research Centre (TS CRC). The Tropical Savannas CRC has been instrumental in facilitating multi-disciplinary research in Australia's tropical savannas and in making knowledge and expertise available to the broad range of stakeholders that use this extensive and increasingly impacted region. In doing so, the TS CRC has played an important role in improving conservation and land management outcomes in the region.

Comments were received on various drafts of chapters in this book from Brydie Hill, Sarah Whiting, and Anna Daniels. Additional comments and assistance on other aspects of the study and publication came from John Woinarski, David Paton, Tom Vigilante, Julian Reid, Peter Dostine, Tony Griffiths, Gabriel Crowley, Jerome Thorbjornsen, Mick Armstrong and Michael Braby. Many thanks to them all.

Jane Bowland patiently tracked down information from disparate sources concerning bustards, most notably including a copy of the original excerpt of Daniel Solander's scientific description of the bustard made available by the trustees of the Natural History Museum, London. Bill Harney, an Aboriginal elder of the Wardaman language group in the Northern Territory, provided valuable insights into the significance of bustards to his people. Thanks also to Tara Dodd of the South Australian Museum for organising the use of the 'Turkey Dreaming' painting by Michael Jupurrulla Rockman and Faye Nungarrayi Gibson. The images of Turkey Flat winery wines were provided by the winery's proprietor, Peter Schulz. The State Library of South Australia provided the image of a bustard shot during the Madigan expedition to the Simpson Desert in 1939. Jim Bannon of the Bustard Eagles Rugby League Club provided an image of the club's crest.

Thanks also to the photographers that greatly enhanced this publication by making their images available: Steve Wilson, Jon Altman, Jonathan DeLaine, Nicole Bartsch, Bruce Doran, Rohan Clarke, James Hager, Robert Elliot, Jamie Anderson and Tony Eales.

Finally and not least, on a personal note to family and friends for invaluable love and support in various guises that made this work possible, thank you to Anna, Alicia, Luiza, Szymon, Brydie, Felicity and Jarrah.

1

TALKING TURKEY

The Australian bustard is an icon of the outback and Australia's heaviest flying bird, yet remains poorly known to most Australians. It is a denizen of open country; of the broad open plains, shrub lands and woodlands that characterise vast expanses of the Australian continent. Having declined from much of southern Australia since European settlement, its contemporary strongholds are in the remote northern and central regions of the country. Travelling these parts and sighting a bustard as it walks imperiously across an outback road is an impressive spectacle and a highlight that makes it into most naturalists' travelogues. Less often witnessed, but even more majestic, are the spectacular display routines that males perform to court females during the breeding season.

One of the most striking physical features of the bustard is its size. It is an impressively large bird. Among the bustard family's greatest claims to fame is that it includes the heaviest of all flying birds. The heaviest recorded weight of a great bustard (at just under 21 kilograms) only marginally eclipses the heaviest Kori bustard ever recorded. By comparison, the heaviest Australian bustard ever recorded was a specimen from Victoria weighing 14.5 kilograms – perhaps not quite competing for the heavyweight championship of the world, but holding the record for Australia's heaviest flying bird nonetheless.

Notwithstanding such notable trivia, the bustard is inherently interesting biologically. It is perhaps surprising therefore that it is relatively poorly known to science. Until recently there had been no detailed studies of the bustard's ecology in the field. Much of what was known was based on descriptive or anecdotal accounts by amateur naturalists and casual observers, while the only detailed study of the bustard's behaviour was based on Kate Fitzherbert's pioneering work on a captive population in southern Australia. One reason for its relative anonymity is its shy, cryptic nature. It generally shuns populated areas, residing mostly in remote regions. It is slow moving, does not fly frequently, and often remains still, relying on camouflage for protection. It is scarce in many parts of its range and has virtually disappeared from much of south-eastern Australia and other settled areas where it was formerly common. For such reasons wild populations are a challenge to study.

First European encounter

During James Cook's first voyage of discovery to the Pacific between 1768 and 1771 he and his crew first landed in Australia at Botany Bay on 29 April 1770. Heading north while charting Australia's east coast in the *HMS Endeavour*, their second landing was at a location near the place Cook called Round Hill Head, near the township known today as Seventeen Seventy. Here Cook went ashore with botanist Joseph Banks and his assistant Daniel Solander, and they became the first Europeans to have sighted an Australian bustard. They promptly shot it. Solander's subsequent description of the collected bird (which given its weight must have been a male) rendered the bustard one of the first scientific descriptions of a native Australian land animal, and the first official technical report of a land animal in Queensland (Figure 1.1). In honour of the first bustard to fall at a European hand the bay in which the *Endeavour* was moored was duly named Bustard Bay.

An excerpt from Cook's journal entry of 23 May 1770 reads:

> *All or most of the same sorts of land and water fowl as we saw at Botany Bay we saw here, besides these Black & white ducks, and*

> *Bustards such as we have in England one of which we killd that weigh'd 17½ (pounds) which occasioned me giving this place the name of Bustard Bay.*

Joseph Banks in his journal made reference to the culinary delights of the shot bird:

> *At Dinner we eat the Bustard we had shot yesterday, it turnd out an excellent bird, far the best we all agreed that we have eat since we left England, and as it weighd 15 pounds our Dinner was not only good but plentyfull.*

Cook's reference to the bustard 'such as we have in England' is to the great bustard; however, by the time he had written these words, great bustard populations across their entire range were already in drastic decline and were very rare in Britain. The primary reason for their demise was hunting, and in smaller part to habitat alteration. But while the birds were hunted for food, it was mostly trophy hunting that was to blame, largely because of the great popularity of firearms, taxidermy and specimen collecting (often under the pretext of science) during this era. As the species declined, so its rarity increased the attraction of collecting it, until eventually

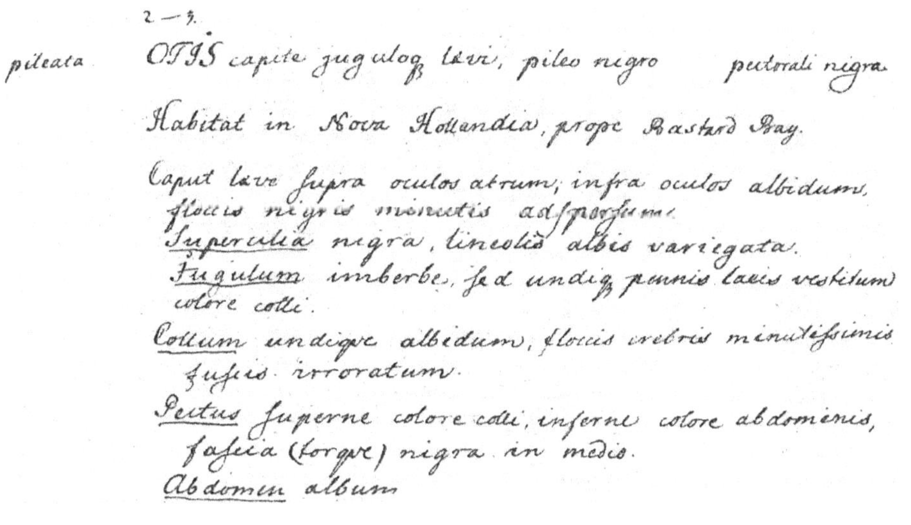
2 — 5.
pileata OTIS capite juguloq; lævi, pileo nigro pectorali nigra
Habitat in Nova Hollandia, prope Bastard Bay.
Caput læve supra oculos atrum; infra oculos albidum, floccis nigris minutis adspersum.
Supercilia nigra, lineolis albis variegata.
Jugulum imberbe, sed undiq; pennis laxis vestitum colore colli.
Collum undique albidum, floccis crebris minutissimis fuscis irroratum.
Pectus superne colore colli, inferne colore abdominis, fascia (torque) nigra in medio.
Abdomen album

Figure 1.1: An extract of Daniel Solander's original description of the Australian bustard as written in Latin was one of the first scientific descriptions of an Australian animal and the first in Queensland. Reproduced with permission of the trustees of the Natural History Museum, London there was no longer a viable population. The last breeding record of the great bustard in the United Kingdom was from Suffolk in 1832.

Place, people and plant names

Bustard Bay was the first of many places named after the Australian bustard in the European lexicon. Just to the north of the bay is Bustard Head (in the Bustard Head Conservation Park). Across the country there is a Bustard Creek, a Bustard Beach, a Bustard Lagoon and a Bustard Spring Gully. There are several Bustard Bores and two Bustard Islands. Near Groote Eylandt in the Northern Territory, there are the Bustard Isles, named by another famous explorer, Matthew Flinders, who passed through the area in January 1803 during his circumnavigation of Australia aboard the *Investigator*.

> There was little wood upon the two sandy isles … they were partly covered with long grass amongst which harboured several bustards, and I called (them) the Bustard Isles.
>
> *Matthew Flinders, 1814*

Across much of outback Australia, the Australian bustard is more colloquially known as the wild, plains or bush turkey – a name owed more to its popularity as an honoured guest at Sunday or Christmas roasts than any close phylogenetic affinity to its North American namesake. This name, sometimes confused with the Australian brush turkey (the mound-building megapode of eastern Australia), gives rise to many more place names. Outside the brush turkey's domain, there is a Turkey Creek in the Northern Territory and in every state except Tasmania. There is a Turkey Bore, Turkey Camp, Turkey Channel, Turkey Cock Gully, Turkey Cock Spring and the Turkey Flat Winery. There are the Turkey Hills, several Turkey Flats, a Turkey Hole, Turkey Heath, Turkey Island, Turkey Lane, Turkey Lagoon, Turkey Plain, Turkey Point, Turkey Rest, Turkey Ridge, Turkey Tank, Turkey Waterhole and a Turkey Well. And when one turkey just isn't enough, there is the Turkey Turkey Waterhole.

There are of course also very many sites of Aboriginal significance named after the bustard in local Aboriginal languages. One anglicised example is Turkey Dreaming, a hill in the Northern Territory, and a sacred site for the local Gagudju people. The Ngarigo people who occupied lands outside of Canberra were themselves named after the bustard which was common on the Monaro Plains prior to the arrival of Europeans.

The bustard also lends its name to common plant names. There are several manifestations of the turkey bush including *Myoporum deserti* and *Grewia retusifolia* (also, and more commonly, known as emu bush). Both plants produce fruits that are a favoured food for both bustards and emus. *Calytrix exstipulata*, a widespread, common shrub in the northern savannas, is also known as turkey bush; however, it does not produce edible parts for birds and its association with turkeys is unclear.

As an aside for the record, the 'turkey nest' in outback Australia is a term used to describe a type of dam that consists of a completely enclosed earth embankment which is filled by pumping water from an alternative source, such as a bore or nearby creek. The term is widely applied across outback Australia, and 'turkey nests' are particularly common on the flat open plains that bustards frequent. However, there is no direct association between the two. The name is derived from the Australian brush turkey and its habit of scraping together earth and leaf litter to form its distinctive nest mound.

Australian folklore

Bustards, while perhaps not as familiar as some Australian animals, have their place in Australian folklore and culture. Like many Australian plants and animals, the bustard has been used as a trademark for business names or logos. One of the first such uses was by the Turkey Flat Vineyard in the Barossa Valley in South Australia. Established in 1847, the winery was named by the original settlers after the large flocks of bustards in the area. Unfortunately, the species is long gone from the region, but it lives on in spirit and is immortalised on the labels of the vineyard's wines (Figure 1.2).

For understandable, if not perhaps somewhat discriminatory reasons, it is unlikely that many Australian sporting teams are ever likely to be named the 'Turkeys' in the ilk of the 'Wallabies', 'Socceroos' or the 'Kookaburras'. But there is at least one sports team that call themselves the 'Bustards' – an over-35s rugby league team in Queensland. The team formed as an offshoot of the Eagle Junction Rugby League Club. While the club's main team retained an eagle in their club logo, the over-35s team adopted their own crest depicting an old eagle on crutches (Figure 1.3). The origin of their name is in fact a clever play on words. As over 35 year

Figure 1.2: The bustard adorns the wine labels of the Turkey Flat Vineyard range.

olds with broken bodies they were in effect the *busted* Eagles but spelt the name 'Bustard' in honour of the bird. Hence, while retaining a hybridised, if not somewhat scientifically suspect, official 'Bustard Eagles' team name, the old boys are affectionately known as 'The Bustards'.

For many people the bustard is often (or at least was) synonymous with food. Almost without exception, the journals of explorers and early settlers that came across bustards, were littered with references to the sport of shooting turkeys or to their gastronomic virtues. Some even offered recipes or tips on how to best cook them.

> *While riding on the plains Dick would sometimes cut down a wild turkey with his stock whip and the old cook would make what he called a sea pie, although a plains pie would be a better title. These turkey pies were a luxury we did not enjoy every day. The wild turkeys, or bustards as they were called, would sooner perish on the*

Figure 1.3: The Bustard-Eagles rugby league club motto.

plain than come into the well for a drink, probably because they knew
what their fate would be if they did.

Tom Booth, Corrong Station, Hay Plain 1870

The bush poet also celebrates the bustard both for its aesthetic attributes
and for its famed eating qualities. The bustard is generally held in high
regard and a fair degree of affection across the outback, and it was without
doubt an important and appreciated source of subsistence for many
outback folk, and in many cases a fondly missed one.

The black scrub turkey's now protected,
like the bustard, grey turkey from the plain.
So we the northern and inland seniors,
won't taste the food from early days again.

Those were hard years, no fancy tucker,
long hours worked, and money always short,
Bird and beast hunted about the country,
game taken for the table never shot for sport.

Young scrub turkey braised with trimmings,
along with potatoes and onions filled the pot.
Whilst the inland fare young plain turkey,
a bushman cooked in camp ovens like as not.

No fancy table settings or crisp white linen,
strong tea, bread or damper, just basic things.
But bush fare slow cooked to perfection,
in those hard years seemed the food of kings.

'The Birds of Black or Grey' – Bernard de Silva

Figure 1.4: An Australian bustard shot during the CT Madigan expedition to the Simpson Desert 1939. Image courtesy of the State Library of South Australia. SLSA: B 20349 – Simpson Desert – 1939

CHORIOTIS AUSTRALIS
(PLAIN TURKEY)

Figure 1.5: Lithograph of the Australian bustard from Gregory Mathews's 'Birds of Australia'.

2

BUSTARD DREAMING

'My Dreaming – Bush Turkey – see the tracks – he goes there and there – all over my country – he makes tracks on this big soakage on my country'

Cowboy Louie Pwerle, Aboriginal elder, Utopia Station

Australian Aboriginal belief systems are based on the Dreaming, a period when ancestral beings inhabited the earth and created the landscape and all its plants and animals. During this time, Aboriginal Law was established bringing meaning and order to the world. According to tradition, laws and beliefs are passed down between generations orally, therefore Aboriginal cultures are rich with stories and legends. Many stories relate to the creation of plants and animals, and often tell of particular physical features, behaviours or other aspects of a species' ecology. Often, these stories also have underlying moral lessons.

Table 2.1: Examples of bustard names in some local Aboriginal languages across Australia

Language	Region	Bustard name
Tiwi	Tiwi Islands, Northern Territory	Kawukawuni
Larrakia	Darwin, Northern Territory	Danimila
Wardaman	Flora River region, Northern Territory	Jegban
Eastern Anmatyerr	Southern Northern Territory	Arwengerrp
Minyerri	Hodgson River region, Northern Territory	Jambirrina
Warlpiri	Tanami Desert, Northern Territory	Wardilyka
Jawoyn	North-east of Katherine, Northern Territory	Penuk
Yulparija	Pilbara, Western Australia	Rankurrji
Gagadja	Great Sandy Desert, Western Australia	Barrulga
Bundjalung	Northern New South Wales	Kaiou gal
Anangu Pitjantjatjara	North-west South Australia	Kipara
Mirning	Nullarbor Plain, Western and South Australia	Idlidja
Ngarrindjeri	Lower Murray Lakes, South Australia	Talkinjeri

The bustard's cultural and spiritual importance to Aboriginal people and its widespread distribution across much of mainland Australia means that it is known to many different Aboriginal groups. Consequently, it goes by many local language names (Table 2.1) and is a subject of numerous Dreaming myths and legends. Across much of Australia, bustards share the same habitats as emus, and with similar ground-dwelling habits and large size, it is not surprising that many stories involve both birds. In some tales the two protagonists are pitted against one another, sometimes with the bustard as the villain. One such story, recounted in several regions and languages, is a tale of deceit that tells how the two birds came to be as they are today.

In the Dreaming, Emu, on account of his large size and the many young he has each year, was regarded as king of all birds. Bustard was jealous of Emu and would watch with envy as Emu would fly high in the air and run swiftly and tirelessly across the plains. Wanting to be king himself he schemed to injure Emu so he could no longer fly, but knowing he was physically outmatched knew he needed to be cunning. Pretending that he had lost his wings, Bustard convinced Emu that all the birds ridiculed him for flying just like they all did, and that to be truly distinguished he should be able to get around by walking, just like bustards and men do. Emu, not wanting to lose face and without suspecting any guile, cut off his wings. When next the two met, Bustard revealed his treachery, and flew off laughing to tell all the other birds of Emu's foolishness. Angry, and now without wings, Emu planned his revenge. Some time later, leaving most of his large brood at home, Emu took his largest two children to feed on the plain where he knew Bustard was with wife and young. Upon meeting the family, Emu saw that the parents were working hard to feed all their offspring. He convinced Bustard that he had too many young, and that he himself had killed all of his others because it was much easier to bring up just one or two strong young like he now had. The following day the two met on the plain again. This time Bustard had only two young, and told Emu that he decided his idea was sensible, and that he had duly got rid of all his other offspring. In response, Emu erupted in laughter, revealed his trick and called Bustard a bigger fool than he had been because a bird's strength lies not in his powers of flight but in the number of his offspring. And so it transpired that emus became flightless and bustards have so few young.

The bustard is also central to a Pitjantjatjara myth from the north-west of South Australia explaining the origins of fire. In the story, *wati kipara* (bush turkey man) was travelling the land while eating *arnguli* (wild plums). He came across some young men with *waru* (fire) and stole it from them. He carried it across the land in his head feathers with the men chasing, trying to snatch the fire away. They gave chase all the way south to Yurkurla (near Yalata in South Australia), all the while trying to snatch the fire away. Reaching the coast kipara ran into the sea holding the fire above his head. Undeterred, the men continued the pursuit into the water, eventually grabbing the fire and spreading it across the land.

Bustards as totem and taboo

Understanding Aboriginal belief systems is fundamental to appreciating the traditional relationship between people, the landscape and the species within it. While in the past technology limited the extent to which certain species could be exploited to a degree, strict laws governed hunting, preventing over-exploitation and ensuring that harvesting was done in an appropriate and sustainable manner. At the base of many of these laws, and by implication incorporating preservation into a moral code, was a system of totems and taboos. Totemism is an integral part of the culture and spirituality of Aboriginal clans, as it forms the basis of kinship systems and signifies the connection between people, ancestral beings and the landscape. Every Aboriginal person is associated with at least one totemic being, but their relationship to their totem may vary. For example, for some groups it may always be prohibited for a person to eat their totem, while for others, the taboo only applies at certain times.

It is common for certain animals and foods to be prohibited for people until they have reached a certain age or level of maturity. Among some Aboriginal groups, for example, it is forbidden for young boys or girls to eat bustard meat or eggs until they have undergone appropriate initiation ceremonies. Should the taboo be broken, the child is told that they may lose their sight or suffer some other unsavoury affliction. Because bustards are an important animal and certain knowledge about them is passed down to worthy initiates only, many of the stories and ceremonies are secret and cannot be made public, either in the Aboriginal or the wider community.

Other taboos may apply to specific sections of the wider community and be generally known. For example, a widely applicable taboo in northern Australia prohibits women from eating bustards while pregnant. Women that break the taboo are said to be at high risk of miscarriage or producing 'mad' children.

For many people, their totem may also be associated with the custodianship of a particular tract of land that they have responsibility for. In such cases the land may be a significant habitat or location for the totemic plant or animal, and the custodian not only has the responsibility to look after the totem directly, but also to manage the land specifically for the totem. People therefore had an intimate and detailed knowledge of the ecology of the flora and fauna on their lands. In some cases, sacred sites may also be areas of high value for totemic species. At such sites these animals cannot be harmed, hence the site may be an important refuge, or in effect, a conservation reserve for the species.

Generally, people for whom a bustard or other animal is a totem may also be charged with retaining specific knowledge of it and the ceremony for that species. Among Wardaman people of the Northern Territory for whom the bustard (called *jegban*) is a totem, the importance of traditional bustard display grounds or 'leks' was explicitly recognised (see Chapter 6 for more on leks). At such sites it was strictly forbidden to hunt large breeding males or females with young during the breeding season. Ceremonial dances and songs for the bustard function to explain the 'law' for the species and serve to pass down knowledge for it (including its place in the landscape and beliefs of the group) to subsequent generations.

Hunting methods

Since its large size makes it a prized bush food, and its elaborately decorated feathers are valued for ceremonies and decoration, the bustard is often hunted. Many people also suggest the type of food bustards eat can taint their flesh. Apparently, bustards that have gorged at mice plagues, for example, have a less palatable flavour, whereas those that have fed on fruit are particularly tasty.

Various strategies were traditionally employed to hunt bustards which generally involved throwing a projectile such as a stick, boomerang or spear. One technique was to exploit the bustard's reliance on camouflage and its tendency to remain still and to hide in thick grass. Along with its somewhat laboured take-offs, a skilful hunter with stealth and a bit of luck could sometimes sneak up on their quarry to within striking range. This technique was particularly effective on hot days when bustards are often reluctant to fly. However, it would rely on first knowing exactly where your quarry was hiding – no mean feat given the bustard's masterful use of camouflage and ability to crawl away on its haunches quietly and unseen.

At other times a hunter would lie in ambush in the grass or in a tree along known trails awaiting a bustard's approach. Fire was frequently used for this purpose to attract bustards into an area. One method was to light a bonfire (rather than a broad-scale bushfire) and use green vegetation to generate a lot of smoke. The hunter would then wait nearby in ambush as a bustard moved in to investigate.

Another technique is related by the ornithologist, AHE Mattingley (1929):

> The best method … (to obtain a close acquaintance of the bird) … is to attract the attention of the bird and keep it fixed on a definite object, such as a kite which is flown over them. Whilst their attention is thus absorbed, a close approach can be made by the observer. The Australian aboriginals, to keep the attention of the Wild Turkey, adopted the stratagem of carrying aloft a fascinating and curious

device, made partly from birds' feathers. This object keeps the birds so interested watching it that the wily aborigine is able either to spear the bird or to throw a noose over its head.

Many Aboriginal groups often use elaborate hand signals to communicate. In some cases sign languages are as well developed as spoken language. One of the most elaborate and best-studied Aboriginal sign languages is that of the Warlpiri people of the Tanami Desert. It is believed to be so well developed because of the tradition that widows do not speak during a mourning period that can last for months or even years when the sole means of communication is by hand signs. Specific signals are also commonly used while hunting game animals, including bustards, when silence and stealth are of the essence. These signals may be variable and will often take different forms in different languages (Figure 2.1).

In many parts of its range, bustards are hunted opportunistically at all times of the year; however, in the northern wet-dry tropics, hunting is limited during the wet season because of restricted access to the preferred habitats of the species. Many Aboriginal people also avoid eating bustards during the wet season because of the apparently high internal parasite loads the birds have at this time. Several people have reported that bustards have worms within the digestive tract and meat. This has not been investigated among Australian bustards as yet; however, helminth parasites

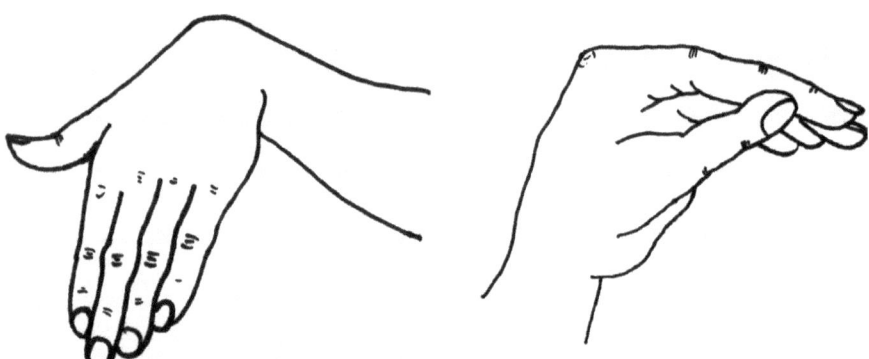

Figure 2.1: Representations of the bustard using hand signals in different Aboriginal languages; (left) the Warlpiri hand sign for the Australian bustard has the hand fully opened and turned face downwards. It is moved sideways at the wrist, in imitation of the movement of the bird's wing. Among several language groups in central Queensland the hand forms the shape of the bustard's head (right). Figures reproduced with modifications from Mountford 1949 and Walsh 1979

Figure 2.2: Bustards are a prized bush food for Aboriginal people. Photo by Jon Altman

are commonly found among other bustard species. A possible explanation for the apparent susceptibility of bustards to parasite and diseases, particularly in the wet season, may relate to the significant weight loss that males undergo over the course of the breeding season. During this time mature males spend considerably more time displaying than foraging. As a result, following peak body mass at the beginning of the breeding season, their mass and body condition decline sharply. By the end of the breeding season males may be emaciated and weak; therefore, their immune systems and bodies are less able to cope with disease and parasites.

3

TAXONOMY AND CHARACTERISTICS

The Bustard is an exquisite fowl
With minimal reason to growl
It escapes what would be
Illegitimacy
By the grace of a fortunate vowel.

George Vaill

For a bird of such a noble, stately appearance, strutting haughtily as it does with bill held high in the air, the bustard's English name does not seem to do it justice. So how did the bustard get its name? Its exact origins are unclear, but go back to antiquity, to Pliny the Roman scholar and historian. In his book *Naturalis Historia* Pliny ascribed the Latin name *avis tarda*, to the bustard (in reference specifically to the great bustard found in his homeland). In modern Spanish the word 'tarda' means 'slow', and it has been suggested that Pliny named the bustard 'slow bird' on account of its leisurely, stolid nature. However, linguists speculate that in old Spanish, *tarda* may share the same origins as the name for bustard in the Germanic languages, namely *trap* in Swedish, *trapp* in Danish, or *Trappe* in German. These words are equivalent to the English 'tread' or 'traipse', and refer to the bustard's long legs and tread. Either way, the Latin name was subsequently translated into *avutarda* in Spanish, *abetarda* in Portuguese, *ottarda* in Italian, *bistarde* in old French, and eventually *bustard* in English.

The Otididae family

The Otididae (bustard) family comprises some 25 species in 11 genera (Table 3.1). It is an Old World family with greatest representation in Africa, with the distribution of several species extending to Europe, Asia and Australasia. Owing to different classification systems that may be based on differences in morphology, behaviour, genetics and ecology, the exact number of species and the inter-relationships between them is subject to continuing debate among taxonomists. The family itself is distinctive and ancient. Its origins are with the Gruiformes, placing them in the same order as cranes, crakes and rails. However, they are considered a separate lineage and are accordingly placed apart in their own infraorder, estimated at having arisen over 70 million years ago.

Bustards are characterised by a distinctive body shape and stately erect posture. They have relatively robust bodies with long thin legs and necks, somewhat large heads with stout straight beaks, and small feet with three forward-pointing toes. Several species exhibit pronounced sexual size dimorphism with males substantially larger and heavier than females. Since all bustards rely on camouflage to differing degrees, plumages are generally cryptic. During the breeding season, males of several species develop elaborate breast or neck feathers that are used for spectacular displays. At this time the motivation to be noticed in order to attract mates counteracts any innate tendencies to otherwise remain discrete. All bustards are exclusively ground-dwelling. They never use trees and nest on the ground. In fact, they are unable to perch given the absence of a hind toe to grip with. Although preferring to walk, they are nevertheless strong fliers. Some species, such as the Houbara bustard, migrate many thousands of kilometres from Eurasia through Central Asia to eastern China and Russia.

Table 3.1: Bustard species and distribution

Common name	Scientific name	Distribution
Australian bustard	*Ardeotis australis*	Australia and southern New Guinea
Kori bustard	*Ardeotis kori*	Southern and eastern Africa
Great Indian bustard	*Ardeotis nigriceps*	West and central India
Arabian bustard	*Ardeotis arabs*	Mauritania to Saudi Arabia
Great bustard	*Otis tarda*	Europe, north Morocco, Central Asia to north-east China
Houbara bustard	*Chlamydotis undulata*	North Africa, Middle-East, Central China, Canary Islands
Ludwig's bustard	*Neotis ludwigii*	Southern Africa
Denham's bustard	*Neotis denhami*	Southern, central Africa to Mauritania
Heuglin's bustard	*Neotis heuglinii*	Horn of Africa to northern Kenya
Nubian bustard	*Neotis nuba*	Sahel zone from Mauritania to eastern Sudan
White-bellied bustard	*Eupodotis senegalensis*	Central, eastern to southern Africa
Blue bustard	*Eupodotis caerulescens*	South Africa
Karoo bustard	*Eupodotis vigorsii*	South Africa
Rüppell's bustard	*Eupodotis rueppellii*	Coastal Angola and north-west Namibia
Little brown bustard	*Eupodotis humilis*	Horn of Africa
Savile's bustard	*Lophotis savilei*	Mauritania to central Sudan
Buff-crested bustard	*Lophotis gindiana*	Horn of Africa, Kenya to central Tanzania
Red-crested bustard	*Lophotis ruficrista*	Southern Africa
Black bustard	*Afrotis afra*	South Africa
White-quilled bustard	*Afrotis afraoides*	Southern Africa
Black-bellied bustard	*Lissotis melanogaster*	Central to southern Africa
Hartlaub's bustard	*Lissotis hartlaubii*	Eastern central Africa
Bengal florican	*Houbaropsis bengalensis*	Southern Nepal, northern India and southern Cambodia
Lesser florican	*Sypheotides indica*	India
Little bustard	*Tetrax tetrax*	Western Europe, northern Africa, central Asia

The Ardeotis genus

The Australian bustard (*Ardeotis australis*) is one of four species that comprise the *Ardeotis* genus, a group defined by their large size and characteristic 'balloon display'. The genus name is derived from the Latin *ardea* meaning 'heron', and *otis*, the Greek for 'bustard', while the species epithet, *australis*, simply means 'southern'. Other members of the genus include the Kori bustard of southern and eastern Africa, the rare and endangered Great Indian bustard of the Indian sub-continent, and the Arabian bustard found in the Sahel region of Africa and the south-western part of the Arabian Peninsula.

Since Daniel Solander's original referral to the Australian bustard simply as *Otis*, the Australian bustard's scientific name has undergone several changes as taxonomists have variously redefined the phylogeny of the bustard group. It was not officially named in the Linnaeus system of binomial nomenclature until Gray gave it the name *Otis australis* (literally 'southern bustard') in 1829, some 60 years after Solander's original description. It was subsequently ascribed to different genera including *Choriotis* and *Eupodotis* before eventually being grouped in its own genus *Ardeotis*. At various times it has been suggested that the four *Ardeotis* species should be considered one species, and for an extended period they were one under the banner of *Ardeotis kori*. Indeed, for many years the holotype (the single physical specimen said to be representative of the species and from which the species is formally described and named) was a Kori bustard from Africa. There are, however, differences in plumage and morphology between the bustards, and several notable differences in display and behaviour.

Australian bustard characteristics

Size

One of the most striking features of the Australian bustard is its large size. The species also exhibits significant sexual size dimorphism, with males much larger than females and on average two to three times heavier. As a comparison, a typical male may stand 110 to 120 centimetres tall and have a wingspan of up to 230 centimetres, while an average female may stand between 80 to 90 centimetres and have a wingspan of around 170 centimetres. See Table 3.2 for more morphological measurements.

Notable among male bustards is the significant variation in size and weight between individuals. Although there is little precise information concerning the lifespan of bustards, they are regarded as a long-lived species, and individuals have lived for over 30 years in captivity. Male bustards appear to continue increasing in adult weight with increasing age, leading to some very large individuals. As noted in the opening chapter, the largest recorded Australian bustard male was an individual shot in Victoria that weighed over 14.5 kilograms, and weights over 10 kilograms were commonly recorded in the early days of settlement. Elsewhere, even larger Kori and great bustards have been recorded. However, such prodigious individuals are no longer recorded and their absence today is likely to be the result of the removal of large males through hunting and the increased pressures facing present-day populations whereby individuals may not survive long enough to reach maximum potential sizes.

There is currently very little information regarding geographical variation in bustard body size and other physical parameters. However, many mammal and bird species exhibit intraspecific tendencies for increasing body size with latitude. In other words, in Australia, the sizes of individuals of the same species tend to be larger in the southern regions than their counterparts in northern Australia. This is a generalisation but it applies to some well-known Australian animals. For example, the largest and heaviest wedge-tailed eagles in Australia are those in Tasmania, while the smallest are those found in northern Australia. The potential implication for bustards, assuming the same geographical patterns in size apply, is that given the widespread and marked declines and extirpation of the species in southern Australia, the large sizes of bustards recorded in the past are indeed that – a thing of the past. Bustards of northern and central Australia, no matter how remote and protected, and no matter how long they live, may never attain the immense sizes of southern males in times gone by. And the ultimate implication of that? The poor old bustard may never again have a shot at the Guinness Book of Records title of world's heaviest flying bird!

Sexual maturity

Another indicator of the long lifespan of bustards is the age at which sexual maturity is reached. Again, the field data are few but observations in captivity demonstrate that bustards exhibit sexual bimaturism, meaning that the sexes mature at different ages. In captivity, females may breed at three to five years, but males may not breed successfully until as old as seven to nine years of age. This significant difference is a result of the specialised mating system that bustards employ whereby sexual selection acts on males by female choice. Since bustards are largely polygynous, with relatively few males doing most of the mating, females are free to choose males with the most desirable characteristics to pass on to their offspring. These males are usually the larger, older and more sexually experienced individuals.

Plumage

In keeping with their cryptic nature, the bustard's plumage is designed to help it blend in with its environment. From a distance it may appear somewhat nondescript. A general impression may be of a large, primarily brown bird with grey-white neck and black crown. However, closer inspection reveals remarkably intricate feather patterns and a range of colour tones.

Male bustards have a black crown, with the elongated black feathers of the nape forming a short crest. By comparison, the crown of females is narrow and brown with a pale tawny or buff patch in the centre. The bustard's chin and ear coverts are white or off-white and a dark grey or black strip of bare skin runs below a white eyebrow from the eye back to the base of the crown. The neck is finely barred with dark grey and white. At a distance this gives the neck a uniform light grey or off-white colour. In females, these patterns are darker but can vary between individuals, giving an overall appearance ranging from a light to a smoky dark grey.

The bustard's dominant body colour is brown. Its back, tail, mantle and scapulars are densely vermiculated with light to dark brown and rufous tones. Separating the base of the neck from the bustard's underparts, and traversing the uppermost part of the breast, is a narrow black band that is thinner or absent in females. The underparts, including the lower breast, belly and thighs, are white or off-white. An eye-catching feature is the black and white chequered patterning of the bend of the bustard's wing which contrasts distinctively with the brown tones of the mantle and back and white underparts.

Perfectly adapted to blend in with their environment, the downy feathered bustard chicks are buff-coloured with brown stripes along the head and neck, and body striped and mottled with brown. Juveniles superficially resemble females but differ most obviously in having off-white to cream spots on their upperparts. At present, sexing juveniles on the basis of plumage is not possible given the limited knowledge of the plumages of young birds.

Table 3.2: Morphological measurements of bustards from the northern tropical savannas (SE = standard error, n = sample size)

	Females				Males			
	Mean	SE	Range	n	Mean	SE	Range	n
Weight (kg)	3.2	0.1	2.4–4.2	14	6.3	0.4	4.3–9.3	17
Head–bill length (mm)	148.6	1.6	138–160	14	185.2	3.6	170–220	19
Beak length (mm)	61.0	2.5	54–68	5	82.5	3.0	75–95	6
Wing length (mm)	463.8	7.1	420–510	14	588.5	9.2	512–640	17
Tail length (mm)	222.4	2.2	200–232	14	262.3	4.6	215–282	18
Tarsus with foot (mm)	162.4	2.1	151–176	14	211.6	2.3	191–225	16
Tarsus (mm)	148.3	2.1	136–156	10	185.3	2.4	174–194	7
Central toe and claw (mm)	58.0	3.0	55–61	2	72.0	5.0	67–77	2

Bare parts

The bustard's sharp, stout off-white to grey coloured bill is an effective, versatile implement for catching and handling a variety of foods. Behind it runs a strip of bare grey skin from the gape back to below the bustard's eye. The eye itself is large and the iris usually a yellow-brown radiating out to a cream colour. The legs and feet of adult bustard's vary from a pale yellow to grey colour, or to sometimes olive tones, which contrast with the pink legs, feet and bill of juveniles.

Vocalisation

Bustards are generally silent. The most notable vocalisations are made by mature males at the height of the display sequence. During display it makes a resonant thudding sound mechanically by closing its bill briefly before emitting a hoarse, guttural call that lasts about a second. This sound has been variously described as a lion-like roar, a snore, and even like a locomotive letting off steam. Despite its low amplitude, it carries a considerable distance, and to the human ear may be heard several hundred metres away. A similar brief hoarse roar is emitted as an alarm call when individuals are flushed. There is little information on the vocalisation of young bustards but an alarm call made in surprise of two short barks has been described as resembling that of a young dog's.

4

STATUS, DISTRIBUTION AND HABITAT

Historical distribution

At various times during the Pleistocene, an epoch characterised by repeated glaciations and spanning approximately 1.8 million to 10 000 years BP, Australia formed a continuous land mass with nearby islands including New Guinea and Tasmania. As sea levels fluctuated, so too did the relative amount of land exposed between the Australian mainland and the surrounding islands. At the height of glacial periods in the north, the present-day Arafura Sea, Gulf of Carpentaria and Torres Strait were mostly dry, forming a more or less contiguous expanse of open grassland plains and woodlands between today's north Australian coastline and southern New Guinea. The name given to this larger continent is Sahul, and the most recent period of connection was some 10 000 to 15 000 years BP.

The bustard's affinity for open grasslands and woodlands, and its widespread distribution today, suggests it would have had a more or less continuous range from the plains at the base of the mountains of southern New Guinea to beyond what are today the cliffs defining the southern edge of Australia on the Nullarbor Plain. As glaciers retreated and sea levels rose, the area of land surrounding Sahul contracted, eventually leading to the separation of the Australian mainland from its northern neighbour. For a while at least, as the sea barrier between the landmasses was not great, bustards may have continued moving between the gradually separating landmasses with the regularity of such movements decreasing as the distance increased. Bustards are capable of strong sustained flight and occasional contemporary records on islands in the Torres Strait suggest that bustards may still sporadically move between New Guinea and northern Australia. However, there is no firm evidence for such movements. Rather, it is more probable that these populations are largely isolated from each other – at least until the next ice age.

At the other end of the continent a similar separation of landmasses occurred between the mainland and Tasmania. However, bustards have not been documented in Tasmania either by early explorers and settlers, or in Aboriginal folklore. The absence of bustards from the island is likely to relate to the ruggedness of its terrain, while its colder climate, which would have been even wetter and cooler during the Ice Age, is less favourable for what is essentially a bird of mild temperate, tropical and arid regions.

At the time of European settlement of Australia, bustards were found almost anywhere that was flat and relatively open. Early explorers and settlers made frequent references to their abundance on their travels. Referring to the 'luxuriant' plains of the Hunter River in New South Wales, and of course without foregoing the opportunity to note the culinary merits of his admired subject, the naval surgeon and settler, Peter Cunningham notes:

> (the Hunter) plains are the great resort of our wild turkeys, which you will see here stalking majestically about, and which afford an excellent and most delicate repast.
>
> PM Cunningham 1827

Bustards were common even on the plains around Melbourne. In the mid-1830s John Batman, the grazier credited with founding the city of

Melbourne, noted 'great numbers' in Williamstown, although he complained they were too timid to allow him to shoot any. Death by rifle, however, was soon to be the fate of many bustards in these more settled regions as the number of settlers increased and cattle were introduced to grassland areas. By the 1850s few bustards were seen around Melbourne, while the influx of settlers and expansion of settlements during the gold rush period ensured they were soon to decline dramatically from suitable habitats in all but the most remote parts of Victoria.

A similar fate befell populations in all other settled regions. From southern Queensland to eastern New South Wales, and from southern South Australia to the goldfields of Western Australia, the decline of bustards was pronounced and continued unabated into the 20th century.

> In Queensland in the late eighties of the last century I have counted ... in a morning's ride ... on the Barcoo, three hundred of these fine birds on country that had been swept the day before by a bushfire ... Those were the good old days for Bustards, however.
>
> FL Berney 1936

> In the last six years that I was in Kalgoorlie (1941–1947), during which time I travelled many thousands of miles of outback roads – I saw only two Bustards in country where they were much more plentiful when I was a child.
>
> HM Wilson (cited in Sedgewick 1954)

Contemporary status and distribution

The bustard remains widespread and locally common in parts of northern and central Australia (Figure 4.1). However, having declined from more settled regions long ago, recent evidence suggests patchy declines in areas that have been considered strongholds for the species. In the Northern Territory, recent declines appear evident from Bird Atlas reporting rates which dropped by 70% between the first (1977–81) and second (1998–2001) atlases – a more substantial decline than for any other state or territory. These declines appear to have been most marked in the southern, arid

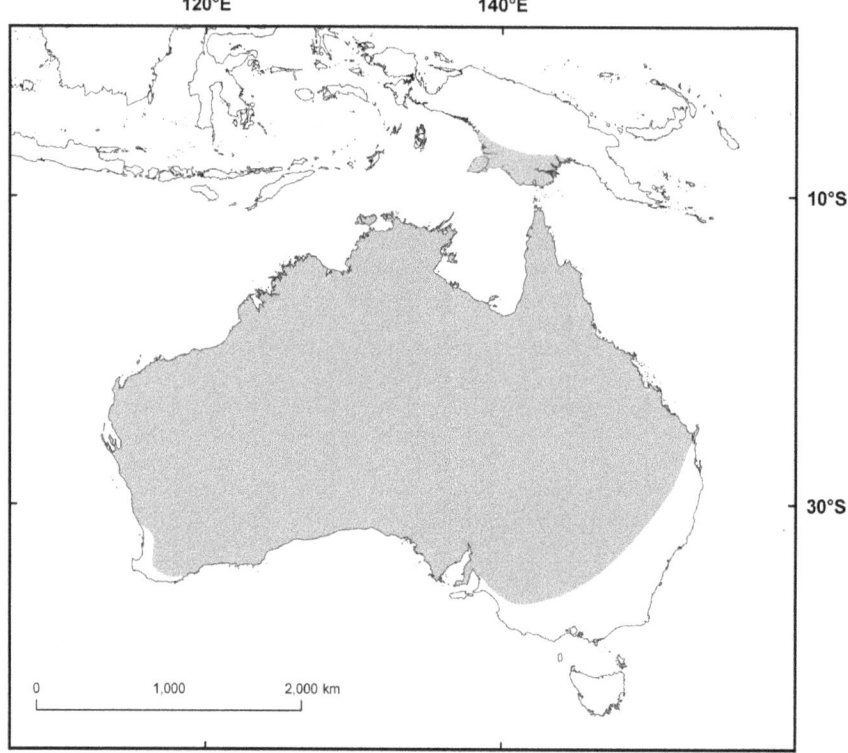

Figure 4.1: Distribution of the Australian bustard in Australia and New Guinea.

regions of the territory, and have led to the recent classification of the species as 'Vulnerable' there. Elsewhere, Aboriginal people and local land holders in remote northern and central regions also report lower numbers in some areas compared to the past. In the Kimberley region of north-western Australia, for example, large flocks of anywhere up to 200 individuals at fires were a reportedly common sight in the past, but such concentrated numbers are very rarely now seen.

Defying the overall downward trend in bustard numbers have been localised increases in some areas. Bustards have seemingly adapted well to recently cleared areas in northern Australia and eastern Queensland. In recently cleared areas such as the Brigalow Belt in Queensland, and horticultural developments in northern Australia, populations are significantly greater than in surrounding 'natural' habitats.

These higher numbers, however, may ultimately reflect immigration or movements from other areas rather than an actual increase in population size. Accurately estimating numbers, population trends and distribution patterns of a species such as the Australian bustard is inherently difficult. Low population densities, the remoteness of much of its range and the apparent tendency for bustards to move in response to changing habitat conditions, result in fluctuations in numbers at any given location that are difficult to monitor. This makes it a challenge to disentangle the extent to which local population changes are a result of movements into and out of a region or real changes in overall numbers that may be caused due to habitat alteration, predators, hunting or other threats, or conversely in this case, favourable conditions in an area.

Habitat preferences

Bustards, as the plains turkey version of their name suggests, have a broad preference for open habitats, ranging from open grassland plains to low shrublands, grassy open woodlands and similar but artificial habitats including croplands, golf fields and airfields. They tend to avoid densely vegetated areas and favour flat terrain over hilly areas. They may also be associated with watercourses, particularly in more arid regions, where broad run-on areas and river channels are often the most productive areas in the surrounding landscape. They also exhibit a strong affinity to fires, and are often seen at fire fronts and in recently burnt areas. At such times they may be observed in large flocks and venture into more densely wooded habitats that fires have opened up.

In the northern tropical savannas, bustards are dependent on a subtle diversity of habitats that satisfy the different seasonal requirements of males and females. In the wet-dry tropics, the seasonal progression from the highly productive wet season (December to March) through the dry season is characterised by a marked reduction in the height and cover of the dominant grass layer as grasses senesce and are progressively removed by fires and grazing. From a bustard's perspective, this presents a highly dynamic seasonal change in habitat conditions that affects both their foraging habits and needs for cover. Since bustards are dependent on their sharp eyesight and camouflage for predator avoidance, they need to be able to both see far and take cover when required. Accordingly, their preferred habitats may be a compromise between the need for effective vigilance and adequate concealment, hence they favour habitats that are neither too sparse nor too dense. Habitats with limited cover provide limited options for hiding, particularly while individuals are roosting in the hotter part of the day or during nesting. Conversely, habitats that are

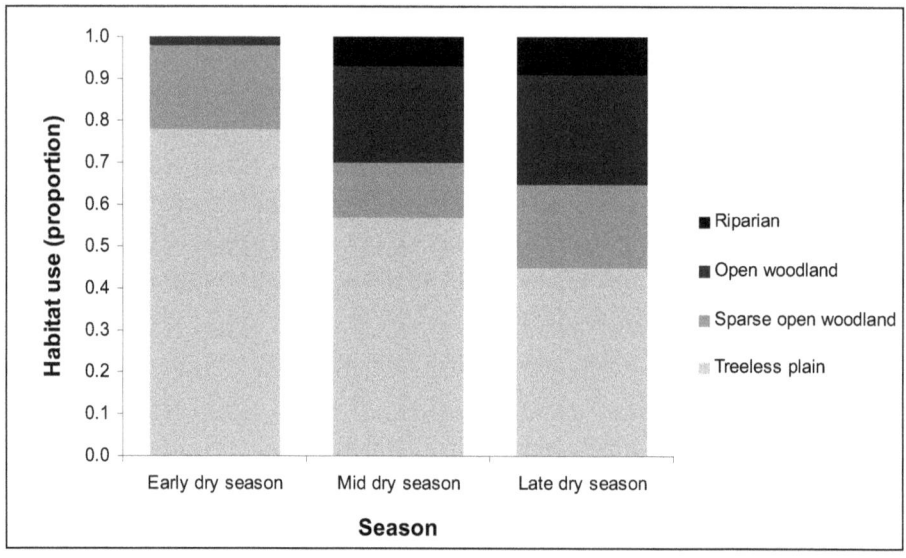

Figure 4.2: General habitat preferences of bustards during the early (April–May), mid (June–August) and late (September–November) dry seasons at Kidman Springs in the northern tropical savannas, expressed as proportion of each habitat used. These data are pooled from observations during systematic vehicle-based surveys of the site each season over two consecutive years.

too dense reduce the ability of individuals to detect and escape from potential dangers by leaving them more open to ambush, and lowering their ability to take flight when required because of their somewhat cumbersome take-offs to get airborne.

During the early dry season in the north (around April and May), when grass height and cover are greatest, bustards concentrate on open treeless grasslands. At this time, these areas are especially rich with grasshoppers, a favoured food, while the thick grass layer provides ample shelter while allowing them to see far by standing or extending their long necks over their grassy substrate. As the dry season progresses the grass layer and food resources decrease. It becomes less advantageous to rely primarily on open treeless plains, hence bustards begin to increasingly use more wooded areas. By the late dry season (September to November), when grass cover and food are at their lowest and temperatures are highest, open woodlands, shrub lands and riparian habitats assume more importance (Figure 4.2).

The late dry season is also the peak breeding period for bustards in northern Australia. Male bustards favour open areas to maximise the

Figure 4.3: An open grassland plain in the Victoria River District, Northern Territory.

conspicuousness of their spectacular display routines, which serve to attract and court females. Females in turn require more sheltered areas with cover for nesting. Among the best areas in the tropical savannas are those where open plains are interspersed with shrublands or open woodland. Broad classifications of bustard habitat associations in the semi-arid regions of the Northern Territory suggest the best quality habitat, supporting the largest concentrations of bustards, are at the margins between tussock grass plains and semi-desert scrub. Similarly, on the Nullarbor Plain in South Australia, it is the interface between low chenopod shrub land and tall mulga shrub land that also support the highest bustard numbers. These ecotones (the transition zones between two adjacent ecosystems), are preferred breeding areas for bustards because they best satisfy the requirements of both males and females.

Horticultural regions

Embedded in the matrix of woodlands and natural grassland landscapes are areas that have been cleared for cultivation of crops. Some of the most

Figure 4.4: Bustards have adapted well to areas cleared for horticultural development in northern Australia.

notable regions in the north include parts of the lower Cape York Peninsula, the Daly River region in the Northern Territory and the Ord River region in the East Kimberley of Western Australia. These areas contrast with surrounding regions on account of their relative stability and productivity, particularly during times when there is low food availability and limited surface water in the surrounding landscape. Irrigated peanut and melon crops in these regions, for example, provide favourable conditions for bustards and an alternative and highly productive food source at a period that is otherwise difficult. Bustards exploit these areas and may be found in numbers significantly higher than surrounding 'natural' habitats.

Patchiness of habitat quality

In Australia's rangelands, variability in climatic conditions, often as a result of highly localised rainfall, largely determines which specific areas may be best for fauna at any particular time. In the wet-dry tropics of the north prior to the onset of broad monsoonal rains, localised storms and rainfall events result in a patchwork of highly productive areas across the landscape. In more arid and semi-arid regions this spatial and temporal variability is more erratic, with patches across the landscape varying unpredictably over longer time periods. Animals that are able to move quickly and efficiently between such patches are able to exploit pulses in food resource productivity opportunistically and with limited competition.

The Australian bustard is well adapted to cope with such conditions by moving readily to favourable habitats opportunistically. On the Barkly Tableland, Max Downes noted the use of such patches in the early wet season. To this end, the occurrence of bustards in particular areas has sometimes been regarded as an indicator of the condition of the landscape. During their ill-fated expedition from southern Australia to the Gulf of Carpentaria, William Wills, of Burke and Wills fame, made mention of the quality of the country they were travelling through on the basis of the presence of bustards.

> *(We) disturbed a fine bustard which was feeding in the long grass; we did not see him until he flew up. I should have mentioned that one flew over our camp last evening in a northerly direction; this speaks well for the country and climate.*
>
> *William Wills 9 January 1861*

a) 'Turkey Dreaming' – painted by Michael Jupurrulla Rockman and Faye Nungarrayi Gibson from Hooker Creek, in the Northern Territory. The painting tells of two bustards that were searching for water through sandhills that took two different routes. One went through the Tanami Desert and found waterholes. The other went through the Kulungalimpa region and also found water. The bustards are represented by foot tracks, waterholes by circles and sandhills by black cigar shapes.

Image of Turkey Dreaming (A69463) provided by the Aboriginal and Archeaology Collections, South Australian Museum, Adelaide, Australia

b) Bustards lay one or two, occasionally three, olive-brown eggs. The nest is a simple scrape on the ground in a bare open area, among grass or below a shrub or tree. Tony Eales

a) A 12-day-old Australian bustard chick. Jonathon DeLaine

b) Female bustard with a juvenile young. Although young leave the nest immediately and develop quickly, they will remain with the female parent for several months and sometimes until the following breeding season.
Jonathon DeLaine

a) Close-up view of a female Australian bustard. Bruce Doran

b) The typical pose of the bustard. Bruce Doran

a) Bustards can go without water for extended periods but will make use of surface water when available. Mark Ziembicki

b) Mature males are distinguished from younger males by their larger size, 'square'-shaped head and more developed throat sac. Mark Ziembicki

a) Masters of camouflage, bustards rely on their cryptic plumage and ability to remain still to blend into their surroundings. Rohan Clarke

b) Bustards have a broad omnivorous diet. This Kori bustard in southern Africa, once considered the same species as the Australian bustard, feeds on a small snake.
James Hager

a) Bustards forage in a range of habitats and circumstances to exploit a variety of foods. During the wet season in northern Australia they will often forage in inundated areas, consuming frogs, freshwater crabs and insects. Jamie Anderson

b) A mature male bustard takes off. Bustards are the heaviest birds that fly, with the heaviest Australian bustard recorded to date weighing 14.5 kilograms.
Rohan Clarke

a) Bustards are strong fliers capable of sustained flight over large distances.
Bruce Doran

b) The spectacular display posture of a mature male bustard at a lek.   Steve Wilson

a) A close-up view of a male bustard's rectrices reveals the fine detailed patterning that decorates its back, tail and mantle. Nicole Bartsch

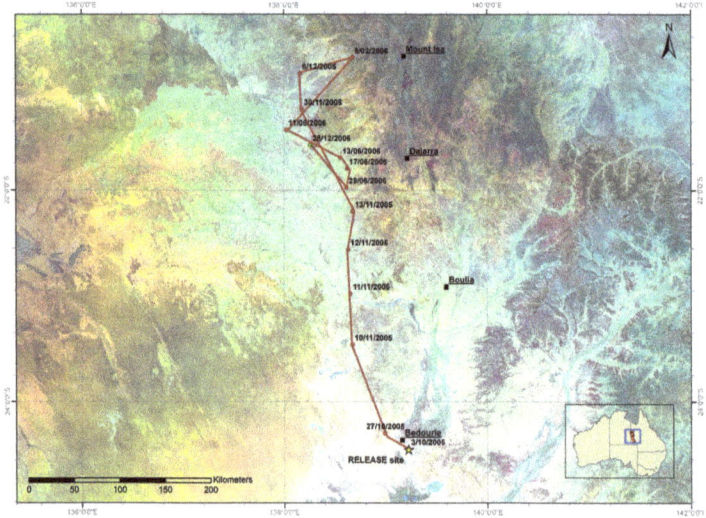

b) The movement path of a female bustard tracked by satellite. This individual was fitted with a solar-powered GPS satellite transmitter near Bedourie in south-west Queensland. Over 15 months it moved north along the Georgina River floodplain to the Barkly Tableland – a distance of almost 500 kilometres – before returning part of the way south and then roaming for several months over a broad, productive region of the extensive floodplain.

5

DIET AND THE DAILY ROUTINE

An omnivorous diet

Bustards will eat almost anything. The list of known food items consumed by bustards is extensive, reflecting a broad, omnivorous diet that includes seeds, fruits, leaves, flowers, green shoots, various invertebrates and small vertebrates (Tables 5.1 and 5.2). Bustards are highly opportunistic, and will gorge on favoured food items when available. To this end they may track or move to outbreaks of preferred foods such as locusts and mice, often into regions beyond their normal range. Their affinity for these foods has even had them hailed as effective biological control agents.

> *The good (bustards) do, especially in the farming districts, is hard to estimate. Grasshoppers are destroyed by them in thousands and*

recently I noticed mice were a favourite item of diet – one bird examined had swallowed twenty mice.

AC Bligh (cited in Berney 1936)

Their opportunistic diet suggests that what they eat is related to the relative availability of food items present in their surroundings. This flexibility in diet stands bustards in good stead to cope with the inherent variability and extreme environmental conditions that characterise a large part of Australia, because it allows them to switch readily between different foods depending on availability. Nevertheless, in spite of their opportunism and tendency to sometimes gorge on single foods, bustards generally have a balanced diet between animal and vegetable components.

There is little quantified information on food preferences and seasonal variation in bustard diets, but a recent study sheds some light on dietary preferences during the mid–late dry season (June to November) in the tropical savannas of northern Australia. The marked seasonality of rainfall that characterises northern Australia drives substantial variation in food resources available to the biota. A short but intense wet season produces a glut of resources that then gradually decline over the extended dry season leading to resource lows by the end of the year. For bustards, the late dry/ early wet season period is an important time of the year because it is the peak breeding period in the north. The study examined the gizzard contents of 33 bustards harvested by Aboriginal people across the Top End of the Northern Territory and northern Kimberley region of Western Australia. Bustards consumed a wide variety of food items but there was an equal balance between insects and the fruits and seeds of a handful of plants (Figure 5.1). Among the insects, grasshoppers and crickets made up the largest mean percentage of gizzard contents by dry weight, with bugs, mantids and beetles also major components of the diet. These foods were also found in the greatest number of gizzards (Figure 5.2). As an example of the bustard's habit of gorging on single foods, several gizzards exclusively contained specific foods, with one individual's gizzard packed with over 140 large grasshoppers.

Diet differences between the sexes

The extreme sexual size dimorphism in bustards, whereby males are notably larger than females, contributes in part to differences in diet between the sexes. Females, with their smaller bodies and bill gapes,

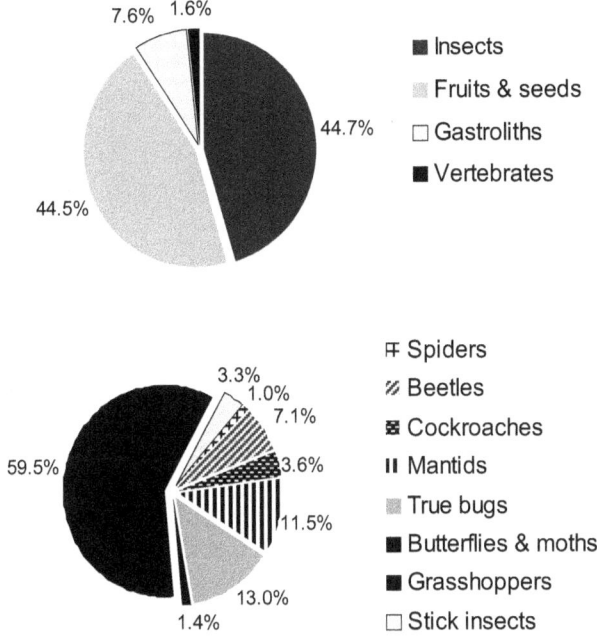

Figure 5.1: Percentages of general food categories (top) and main arthropod taxa (bottom) consumed by bustards in the northern tropical savannas during the mid to late dry season (June to November). Percentages are means of each food item per gizzard measured by dry weight present in 33 gizzards.

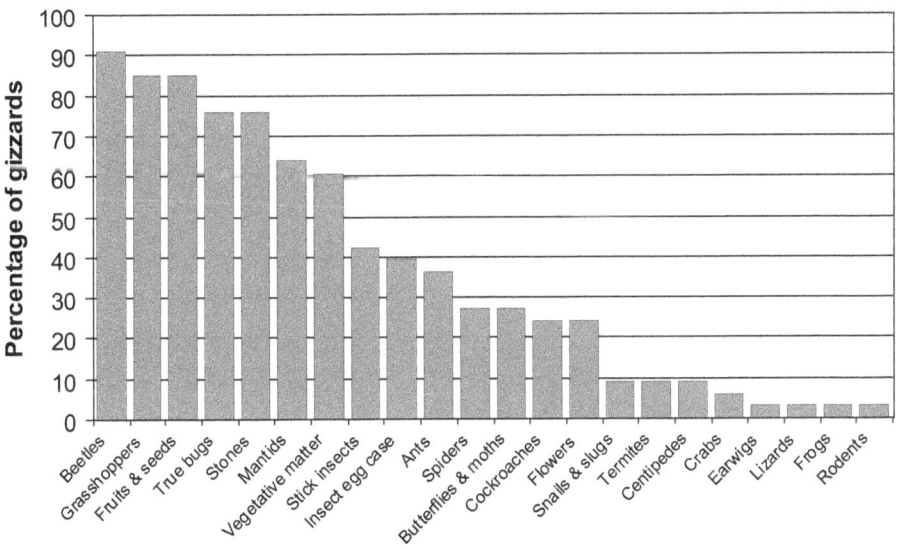

Figure 5.2: Percentage of gizzards (from a sample of 33) containing major food items consumed by bustards in the northern tropical savannas in the mid to late dry season (June to November).

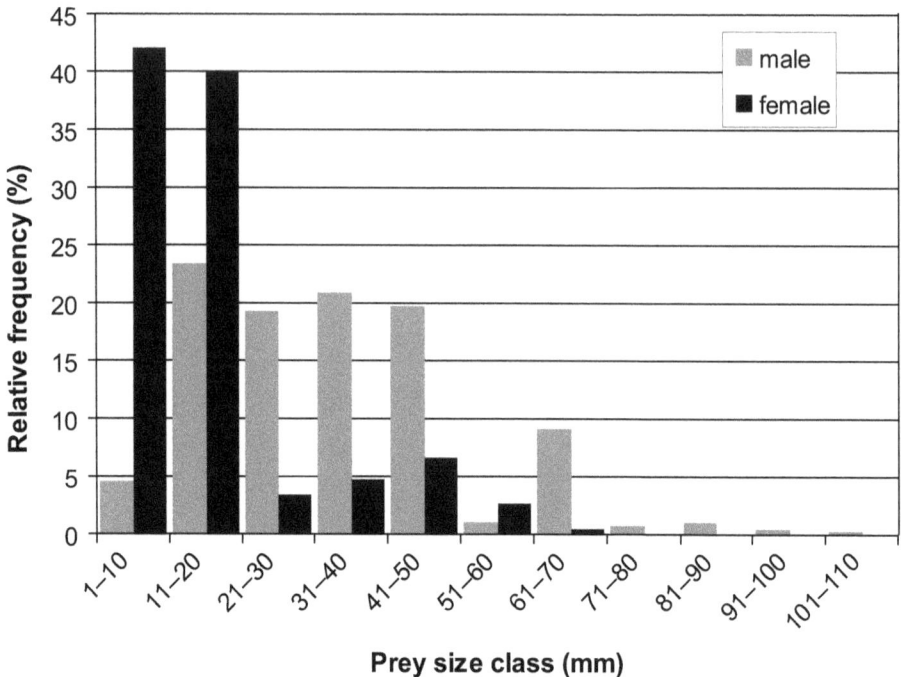

Figure 5.3: Relative frequencies of different invertebrate prey size lengths recorded from a sample of 33 bustard gizzards in the northern tropical savannas in the mid to late dry season (June to November).

generally eat smaller prey and less food by volume. Perhaps somewhat surprisingly for such a large bird, bustards consume a large proportion of rather small invertebrate prey. Among a sample of females in the aforementioned study of diet in northern Australia, 82% of prey items consumed were less than 2 centimetres in length with many items measuring less than a centimetre. By comparison, among males the percentage of such small prey items was under 28% (Figure 5.3). Females took relatively few large prey items: less than 15% were larger than 3 centimetres, whereas for males the proportion of prey over 3 centimetres was more than half, with some large items including agamid lizards of over 35 centimetres in length.

Foraging behaviour

The bustard's sharp, straight, stout bill is all about versatility. It enables them to grasp, probe, capture and manipulate a large range of food items.

They also have excellent eyesight that is used to great effect to glean insects that rely on camouflage from vegetation. Bustards can be surprisingly agile. They may chase down prey that scamper across the ground or jump into the air to snatch at an escaping grasshopper. But more often than not they take a more measured, leisurely approach to foraging. Feeding bouts usually consist of walking slowly with head and bill pointed slightly downwards. Small to medium-sized insects are seized and held in the bill before being swallowed whole. Larger items may be held in the bill and killed or broken up by being hit against the ground before being swallowed. Upon encountering food items such as fruits, buds or leaves a series of short, repeated pecks at the item are made.

Bustards are well adapted to arid conditions and may be able to go for extended periods without directly drinking by gaining water from the vegetable matter and other foods they eat. They will, however, make regular use of surface water if available, particularly during dry, hot spells. During the late dry season in the northern tropical savannas, bustards will regularly move to drink at waterholes in the morning and late afternoon. To drink, they may stand or sit on their tarsi and take water into the mouth and swallow it by lifting the head and tilting it back.

Bustards are attracted to fires where they chase prey escaping the fire front or collect foods made accessible by the fire, including dead insects and small vertebrates found in recently burnt areas. In this context, it may be said that bustards eat carrion but the habit has not otherwise been recorded. However, among some bustard species in Africa, including the closely related Kori bustard, individuals have been observed carrying and consuming the corpses of small birds killed along roadsides.

The large quantities of fruits and seeds bustards eat mean they are important plant dispersers. Plant species such as turkey bush (*Grewia retusifolia*), melons and the various *Cassytha* species with their hard seeds are likely to benefit from their attractiveness to bustards. It has even been proposed that bustards (or an ancestor) may have played a possible part in the dispersal of a lineage of melon of the *Austrobryonia* genus from south-east Asia to Australia in the past.

Unusual foods

With such a broad diet bustards invariably consume some rare and unusual foods. Among the more curious food items consumed is the inland crab *Holthusiana transversa*. These crabs are distributed across parts of northern and arid Australia in temporary watercourses and waterholes. In the Victoria River District of the Northern Territory, they are found on alluvial black soil plains that are waterlogged or inundated during the wet season. During dry periods the crabs aestivate in burrows below the surface but emerge following rains sometimes to become the prey of bustards.

On the aquatic theme, bustards also consume amphibians opportunistically and are sometimes observed foraging thigh-deep in swampy areas when they may take frogs on emergent vegetation. Recently, the distribution of cane toads has spread further into the bustard's northern strongholds. Cane toads, having been introduced to Australia, have several adverse effects on native species. Their voracious appetites for insects and other invertebrates, and their sheer abundance (particularly along advancing fronts colonising new areas) means they consume large quantities of these foods. This not only reduces the abundance of prey species, but in doing so represents serious competition for native species that rely on the same foods. However, the most tangible impact on native species results from the toad's toxicity. The ingestion of a toad, or sometimes just some of its toxin, is enough to kill many native predators. Some of the most susceptible animals include goannas, several snake species and mammals such as the northern quoll. Their impacts on birds is as yet less clear. However, there is anecdotal evidence that several species, including the bustard, can eat cane toads apparently without ill effect. Several observations of bustards eating whole cane toads have been reported from horticultural areas in northern Australia, while in the Gulf of Carpentaria region the gizzards of bustards harvested by local Aboriginal people are sometimes full of young toads.

> *The bustard, which abounded in the Riverina and on the north-west plains of New South Wales, also in Queensland before the rabbits, must, because of its numbers and liking for small snakes, have killed more than any other bird. It may have been the wiping-out of the 'turkey' by poison and by shooting in the north-west that allowed death-adders to increase to frightening numbers between 1910 and 1920 ...*
>
> E Rolls 1969

Bustards have also been hailed as master snake killers to rival other birds. However, this may be somewhat of an exaggeration, since while snakes are eaten and are documented as a common food among other *Ardeotis* bustard species, the consumption of snakes by the Australian bustard is surprisingly rarely noted in the formal literature.

Another somewhat unusual food of the bustard is the sap from trees and shrubs. This food source is likely to be more common than reported, and is known about among Aboriginal people. In the Sahel and other parts of Africa, the closely related Arabian bustard is regarded as a pest in some gum-growing regions because of its appetite for gum which the local human population harvest. For this reason, in some parts the Kori bustard is called 'Gompou' – meaning gum-eating bird.

Completing the extraordinarily broad range of items recovered from the gizzards of bustards are children's marbles. Several bustard species in Africa have also been recorded taking coloured beads. It is uncertain whether these items are confused with food items such as fruits or seeds, or possibly colourful beetles, or whether they are taken as gastroliths. A gastrolith, as the roots of the word imply, is a stone, pebble or grit that may be deliberately ingested to aid in digestion and the mechanical breakdown of food. Of the 33 gizzards examined in the aforementioned study of bustard diet in the tropical savannas almost 80% had some form of gastrolith. These items are also commonly found in a range of other animals including herbivorous birds, crocodiles, seals and, prehistorically, among dinosaurs.

Bustard behaviour

Like most diurnal birds, bustards have a bimodal pattern of activity to their day. They are active from first light for two to three hours, rest for much of the middle of the day, then resume more active behaviour in the late afternoon which continues until after sunset. These patterns of activity are primarily a response to daily temperatures whereby individuals avoid activity at the hottest times to avoid heat stress and excessive water loss. In cooler regions or during cool or overcast periods, the amount of activity in the middle parts of the day may be higher.

Typically, bustards spend most of their active time foraging. When not looking for food they are generally resting, which may include sitting hidden in grass or standing beneath a shrub or tree. While resting in this way in the middle parts of the day, bustards will intermittently commence walking slowly for five to 10 minutes to another location before again settling down to rest for 20 to 30 minutes. They will repeat this throughout

Table 5.1: Animal food items recorded in the diet of the Australian bustard

Class	Order	Family	Species	Common name
Arachnida	Araneae			Spiders
	Scorpiones			Scorpions
Crustacea	Decapoda	Sundatelphusidae	Holthusiana transversa	Inland crab
Gastropoda	Pulmonata			Snails and slugs (incl. egg masses)
Insecta	Coleoptera	Carabidae		Ground and tiger beetles
		Cerambycidae		Longhorn beetles
		Coccinellidae		Lady beetles
		Curculionidae		Weevils
		Dytiscidae		Water beetles
		Elateridae		Click beetles
		Scarabaeidae		Scarab beetles
		Tenebrionidae		Darkling beetles
	Dictyoptera	Blattodeae		Cockroaches
		Mantidae		Preying mantis (incl. egg cases)
	Hemiptera	Cicadoidea		Cicadas
		Pentatomidae		Stink bugs
		Reduviidae		Assassin bugs
	Hymenoptera	Formicidae		Ants
		Vespidae		Paper and potter wasps
	.	Tiphiidae		Flower wasps
	Lepidoptera			Moths, butterflies, larvae
	Orthoptera	Acrididae		Short-horned grasshoppers

Class	Order	Family	Species	Common name
		Gryllidae		True crickets
		Tettigoniidae		Katydids
	Isoptera			Termites
	Phasmida			Stick and leaf insects
	Dermaptera			Earwigs
Myriapoda	Chilopoda			Centipedes
	Diplopoda			Millipedes
Reptilia	Squamata	Agamidae		Dragon lizards
		Scincidae		Skinks
		Serpentes (sub-order)		Snakes
Mammalia	Rodentia	Muridae	*Mus musculus*	House mouse
Amphibia	Anura	Bufonidae	*Chaunus (Bufo) marinus*	Cane toad
				Frogs
Aves				Small birds, quail & nestlings

Table 5.2: Plant foods recorded in the diet of the Australian bustard

Family	Species	Common names	Food item
Apocynaceae	*Carissa lanceolata*	Conkerberry	Fruits
Asteraceae	*Taraxacum officinale*	Dandelion	Flowers
	Blumea spp.	Daisies	Flowers
Capparaceae	*Capparis spinosa*	Spineless caper	Fruits
Chenopodiaceae	*Atriplex* spp.	Saltbushes	Shoots
Cucurbitaceae	*Citrullus colocythis*	Bitter cucumber	Fruits/seeds
	Cucumis melo	Melon	Fruits/seeds
	Cucumis myriocarpus	Paddy melon	Seeds/leaves
Cyperaceae		Sedges	Flowers/leaves
Fabaceae		Legumes	Fruits/seeds/leaves
Haemodoraceae	*Haemodorum brevicaule*	Bloodroot	Flowers
Lauraceae	*Cassytha* spp.	Laurel vine	Fruits/seeds
Mimosaceae	*Vachellia (Acacia) farnesiana*	Prickly acacia	Sap
Myoporaceae	*Myoporum deserti*	Turkey bush	Fruits/seeds
	Eremophila spp.	Turkey, or emu bush	Fruits/seeds
Poaceae		Grasses	Seeds/leaves
Polygonacaea	*Emex australis*	Three-cornered jack	Seeds
Portulacaceae	*Calandrinia* sp.	Parakeelyas	Flowers
Solanaceae	*Lycium ferocissimum*	African boxthorn	Fruits/seeds
Tiliaceae	*Grewia retusifolia*	Emu bush, Turkey bush	Fruits/seeds
Zygophyllaceae	*Nitraria schoberi*	Nitre bush	Fruits

the middle part of the day. If in a small group, individuals will often repeat this behaviour in synchrony and it is a curious sight to see three to four individuals separated by up to 100 metres all rise more or less in unison and begin moving before settling again. The exact reasons for this behaviour are unknown but may be related to predator avoidance.

Seasonal variations in behaviour

The northern wet-dry tropics of Australia are characterised by marked seasonal fluctuations in food resources for bustards. The wet and early dry seasons are periods of high food abundance. Food availability subsequently gradually declines leading to resource lows and a period of hardship towards the end of the dry season. Intuitively, it may be expected that less time is spent foraging when resources are abundant in order to meet energy and nutritional demands than when resources are low. However, among male bustards the opposite is the case. Males spend considerably more time foraging in the early to mid dry season than in the late dry season (Figure 5.4). The reason for this somewhat counter-intuitive circumstance relates to the mating system of the species in the north. The late dry season signifies the onset of the breeding season when mature males commence displaying to attract mates. They subsequently display for extended periods at the expense of foraging. Consequently, during the non-breeding season males must forage intensively in order to accumulate large fat reserves in preparation for a relatively long breeding season when efforts are devoted to attracting mates. Studies of bustards in captivity corroborate observations in the field. During a study of captive birds in southern Australia males spent substantially more time displaying during the breeding season, sometimes almost to the complete exclusion of foraging, thereby significantly affecting their weight and body condition. The weights of males peaked just prior to the start of breeding then declined sharply during the breeding season. Following the cessation of displaying, males gained weight gradually until the following breeding season when the cycle began again.

Subordinate males spend significantly less, if any, time displaying compared to dominant males. They are therefore free to forage for longer periods and hence rely less on accumulated fat reserves. As these males mature, their foraging strategies, and associated patterns of seasonal weight variation, more closely resemble those of larger males. Indeed, as males grow they devote progressively more time to display. By comparison, females spend comparable amounts of time foraging between seasons.

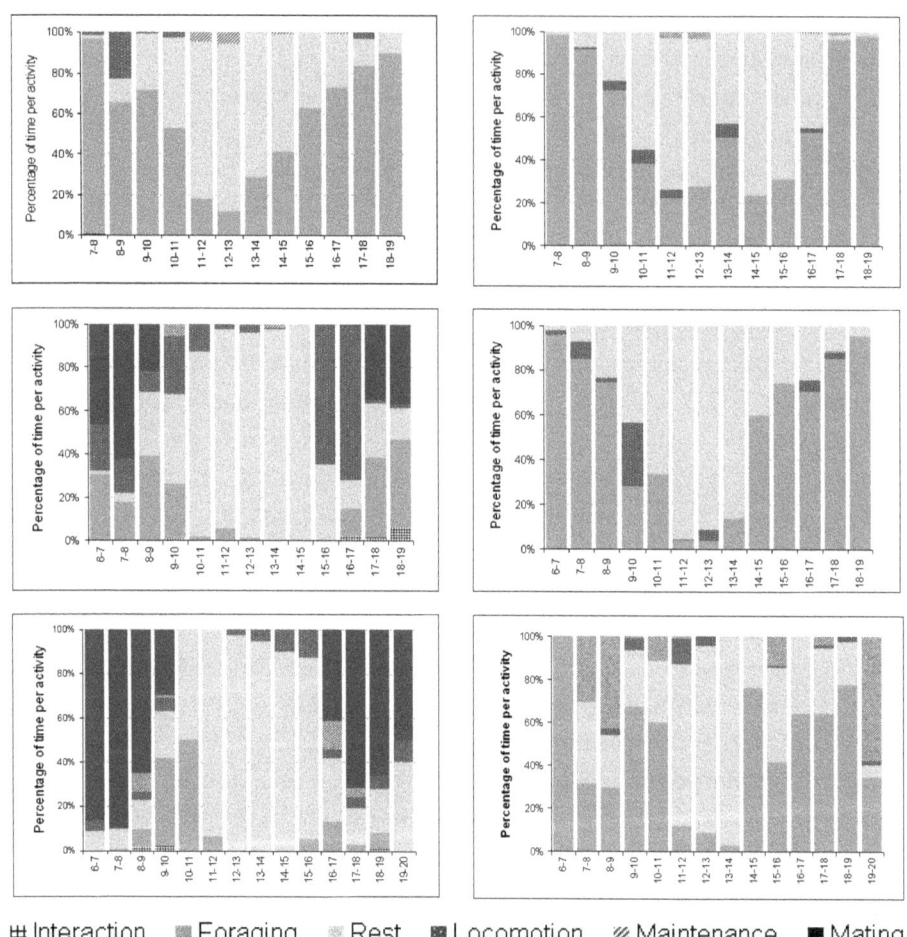

Interaction ░ Foraging ░ Rest ▓ Locomotion ▨ Maintenance ▪ Mating

Figure 5.4: Activity patterns of male (left) and female (right) bustards during the early dry season (top), late dry season (middle) and wet season (bottom) in the tropical savannas of northern Australia.

Other behaviours that change in frequency between seasons include the amount of time spent in locomotion and preening. During the late dry season, when temperatures and humidity levels are high, bustards generally spend more time in locomotion because of their preference for walking rather than flying between roosting and foraging sites. Preening is more pronounced during the wet season, particularly following rainfall.

Social behaviour

Bustards are usually at least loosely gregarious. In the northern savannas group sizes vary between seasons with the frequency of larger groups

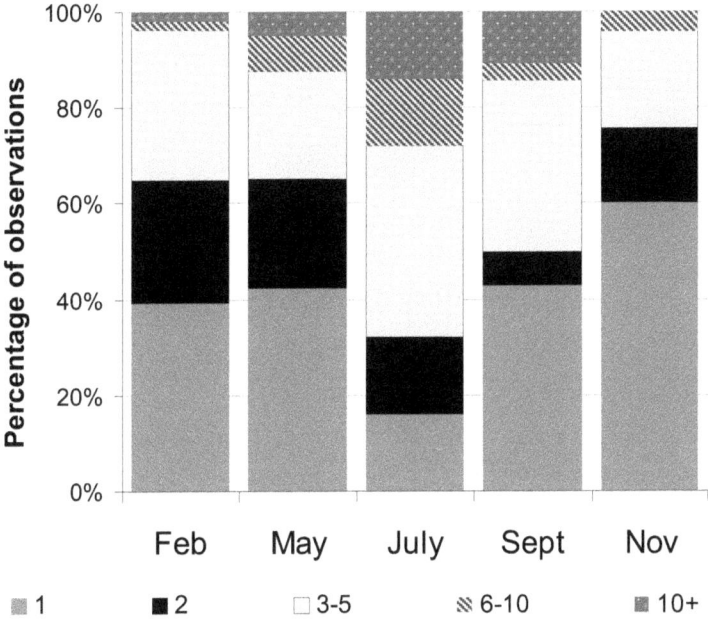

Figure 5.5: The frequency of bustard group size observations varies between seasons in the northern tropical savannas with larger groups most commonly observed in the non-breeding season in the middle of the year. During the breeding season females and mature males tend to be more solitary.

highest in the middle of the year in the non-breeding season (Figure 5.5). At this time groups may commonly range between two to five individuals with occasionally larger groups. During the breeding season mature males and females tend to become more solitary. Groups tend to segregate according to sex but may come together in loose flocks to forage. Small family groups may also be seen, often consisting of female with young.

Larger numbers of bustards often aggregate at abundant food resources and in recently burnt areas. Some flocks recorded in the past have numbered in the many hundreds, but such groups are rarely seen now, with numbers of 50 to 100 together at once a noteworthy sight.

6
EXPLODING BUSTARDS

Although the Australian bustard aspires to the title of heaviest flying bird, such trivia pales into insignificance alongside an altogether more remarkable feature of the bustard's biology – the tendency for males to explode when breeding! Most bustard species appear to be polygynous, meaning that a male may mate with several females, and many species do this within the framework of a 'lek' mating system. A lek is broadly defined as a communal display ground or arena where males aggregate to attract and court females, and to which females come for mating. The word is derived from the Swedish verb *leka* meaning 'to flirt in a playful manner'. Lekking is a specialised mating system that generally favours a select few, particularly fit individuals. Females choose the male they judge to be the best and fittest on the basis of characteristics such as their display, plumage, size, and their geographical position in the lek.

There are several features that characterise 'classical' lekking species:

- males do not contribute to the care of young;
- there is at least some degree of aggregation of displaying males on leks;
- females encounter no other resource on the lek other than males; and
- females are free to choose their mates.

In addition, other auxiliary characteristics may be common to lekking species or populations. Lekking species often exhibit pronounced sexual dimorphism (e.g. plumage or size differences between the sexes), and sexual bimaturism (one sex – usually the female among bustards – reaches sexual maturity sooner than the other). Also typical among such species are the ritualised displays of males and the use of traditional lek sites between seasons and generations.

Leks are described as 'exploded' when males within a display aggregation are separated by considerable distances (up to 1 to 2 kilometres) and it is only when displaying males are mapped over larger scales that such aggregations (leks) are apparent. Exploded leks do not strictly fit the 'classical' lek mould because males may hold territories within which females may forage and nest, potentially violating the third characteristic listed above. Where these male display territories hold critical resources for females that are defended they are said to be resource-based leks. Exploded lekking occupies a position between classical and resource-based leks and appears to be the strategy mostly, but as detailed below not exclusively, exhibited by the Australian bustard.

Lek configurations and social structure

The degree of aggregation of displaying male bustards on breeding grounds is relatively fluid and may range in a continuum from solitary displaying males through more dispersed (exploded) aggregations to relatively tight clusters of several individuals separated by as little as 40 to 50 metres. Such tightly clumped aggregations more closely resemble 'classical' leks in so far as the males are in close, visual contact and there are no resources for females on the lek other than the males (or more specifically, their genes). Aggregation patterns may differ between locations and may change over the course of a breeding season at a given site and appear largely as the result of population densities and prevailing environmental conditions.

A typical scenario in the tropical savannas of northern Australia (the bustard's main contemporary breeding range) is as follows. During the non-breeding season, bustards tend to associate in groups, with small all-male groups a common sight. Just prior to the onset of breeding, mature males begin to disperse to traditional display sites and males may be observed displaying alone intermittently. As the breeding season progresses, more males start to display, and time spent displaying by individuals also increases. At this stage males conform to an exploded or dispersed aggregation (lek) whereby they may be separated by considerable distances (up to a kilometre or two but usually within at least hearing distance of each other). Such aggregations may only be discernible if the distribution of displaying males is mapped over a relatively large scale. The onset of the first rains in the late dry (September to November), which by their nature may be highly patchy in their spatial distribution, stimulates further activity. At the peak of the lekking period, when there are the greatest numbers of displaying individuals, males may aggregate in concentrated clusters of up to half a dozen or more individuals. At this time other males may either continue displaying concurrently at their established sites or join the clusters; the dynamics of how these aggregation patterns operate remain poorly understood. As the peak period of display, which may only last a few weeks, passes clusters disband but dispersed individuals continue displaying at established sites. In at least some parts of the north males may continue displaying in dispersed groups or alone until the late wet season in March.

Dominance hierarchies

The basis of the lek mating system is competition between males for females based on an established social hierarchy. Females choose the fittest males based on a variety of potential characteristics including physical size, plumage, length and vigour of display, and spatial position in the lek. The relative importance of these characteristics in determining female choice are not well known for the Australian bustard, although they may be inferred from anecdotal observations and from knowledge of other lekking species. In general, it is the largest males, with the most spectacular plumage, that display vigorously for the longest period that are the fittest and most attractive to females. Among several lekking species, males at the geographical centre of the lek are also the dominant individuals.

The dominance hierarchy among males is established through intimidation, brinkmanship or outright aggression. Aggressive encounters or fighting mostly occur between subordinate males establishing

Figure 6.1: Males 'squaring-off' in a ritualised bout to establish social dominance and access to females. Photo by Bruce Doran

dominance status amongst themselves. In extreme cases such interactions may lead to blood-stained plumage and injuries. Large, dominant males suppress attempts by potential rival subordinate males implicitly, through intimidation on account of their size or, if need be, by aggressively chasing subordinates away from their display areas. Physical aggression between mature, equally matched rivals is less common. Instead, similarly sized males will often 'square-off' against one another through a ritualised bout. The sequence appears almost choreographed. A male will approach another on his display site and the two begin walking in tandem, following and matching each other turn for turn while separated by no more than 2 to 3 metres (Figure 6.1). These bouts, which are conducted while maintaining the display posture (described in detail in the following section), may last up to half an hour or more, and cover several hundred metres. Eventually, one male relents and presumably loses his position in the hierarchy, or fails in his attempt to usurp the other.

Although dominant males do most of the mating in lekking populations, subordinate males may also occasionally and opportunistically acquire mating chances. Subordinate males may linger on the fringes of leks or they may opportunistically display at suboptimal periods when females are less inclined to nest, for example, early or late in the breeding season.

Lek and display site fidelity

Characteristic of many lekking species is the use of traditional lek sites that are consistently used between seasons and generations. In northern Australia bustards exhibit strong fidelity to leks and males will use the same specific display sites repeatedly within and between seasons as long as habitat conditions at sites remain favourable. In these monsoonal regions, marked seasonal differences in primary productivity, particularly the change in height and cover of the grass layer, lead to significant changes in habitat suitability for bustards between seasons and years. If grass cover increases on and around display sites, visibility may be significantly reduced rendering sites unsuitable for display and leading to their abandonment.

This dynamism recasts the suitability of display sites across the lek within and between seasons. Consequently, it has important implications for the size and configuration of leks and by implication, their social structure. In classical leks, a male's geographical position is often governed by prevailing dominance hierarchies between males whereby the most dominant and successful males are located in the best display positions, often in the centre of the lek. Continued competition for favoured sites naturally results in turnover of males between specific display sites. But how do such frequently changing habitat conditions affect the social dynamics of leks among bustards? At present we do not know enough about the social structure of leks to be able to say. However, one possibility is that stable sites that tend to remain consistently favourable to displaying males within and between seasons (e.g. stony areas with no grass growth) are at a premium and are the most preferred or lucrative sites. These sites may therefore be occupied by dominant males. In contrast, those sites that are more dynamic both within and between seasons, tend to be used more opportunistically by floaters, or subordinate males that do not hold permanent display sites. It may be that instability in habitat suitability may be more favourable to these less dominant males than would be the case if the leks were more permanently fixed with well-established social structures.

Flexibility in mating strategies

Many bird species, including several bustard species, are known to exhibit pronounced intra-specific variation in the adoption of particular mating strategies. Such variation may depend on factors such as resource distribution, population density, female dispersal and habitat stability, such that a species, population or even individual, may exhibit different strategies across its range at different times. As discussed, bustards in the northern savannas appear to exhibit characteristics consistent with exploded and classical lekking, as well as undertaking solitary display. Dynamic habitat and environmental conditions are even more pronounced in arid zones, where the variability between periods is less predictable. It is not known whether bustards form leks in these more climatically unpredictable regions, or revert to some other mating strategy. Patchy and unpredictable resource distribution and low population densities of bustards are a feature of these more arid parts of Australia. In these areas it is unlikely that large numbers of bustards aggregate on lekking grounds (at least not on a regular seasonal basis); rather, they may be more opportunistic in their breeding strategies, adopting other systems, possibly even monogamy. Indeed, in such areas bustards may be forced to move to more favourable regions or go for extended periods without breeding. The breeding of Kori bustards in arid regions of southern Africa, for example, is closely tied with rainfall and may be greatly reduced or not occur at all in drought years. The ephemeral nature and flexibility of lekking behaviour has been emphasised as an important adaptation to variable environments. Among Australian bustard populations, such flexibility may be another example of the adaptation of the species to highly variable environments.

A ritualised display routine

Regardless of the mating strategy adopted, males undertake spectacular displays to attract and court females that are a sight to behold but one that relatively few people get to see. The display sequence comprises several stages (Figure 6.2). At the beginning of a display bout the male stands in position with body and head tilted slightly backwards. He begins to inflate his elaborate, plume-covered throat sac, which gradually extends from the breast downwards until it brushes the ground. The head is then tilted further back and the bill pointed into the air while the feathers of the throat are raised by the inflation of cheek pouches. To complete the display posture he arches his tail feathers backwards so that their tips touch the lower base of his neck. By doing so he also exposes the white tips of his tail feathers further increasing his visibility. This position may then be held or the male may begin rotating slowly or walking in tight circles while swaying the extended throat sac from side to side. At this stage he may also commence his calling sequence, which consists of a short, low booming

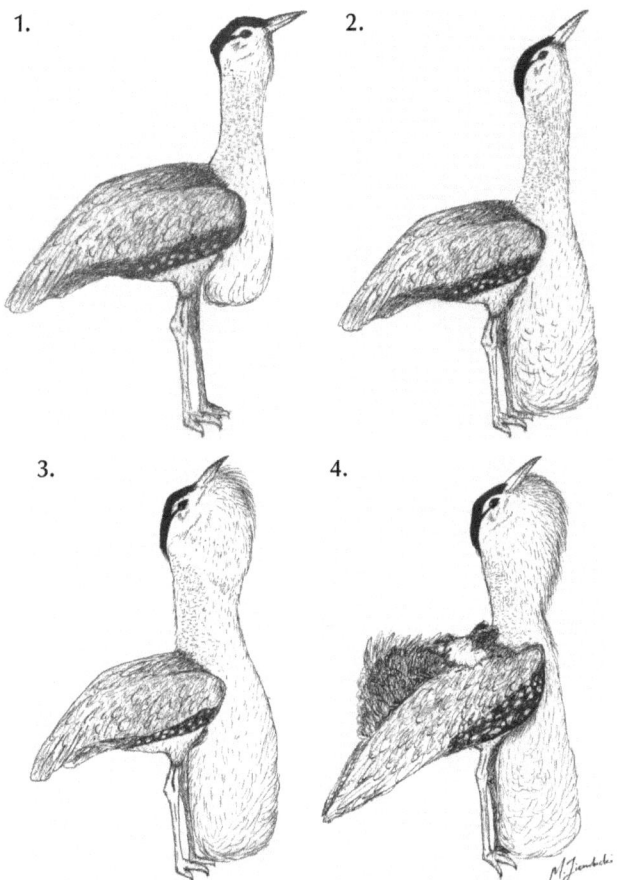

Figure 6.2: Display sequence of the male Australian bustard.

vocalisation, described as a lion-like roar, which is produced every 10 to 15 seconds. This far-carrying call is highly effective in the open grassland and woodland habitats favoured by bustards, which combined with the male's large size and inflated throat sac ensures that he can be seen and heard for a considerable distance.

Upon the approach of a female the intensity of the male's display increases further. He will leave the display area and follow the female, staying within 2 to 3 metres of her while maintaining the display posture. At intervals he may pause momentarily to turn side on to the female to call and then resume his pursuit. This sequence may last for up to half an hour. If the female is receptive she will sit and the male then circles her repeatedly. He positions behind her, and increasingly steps on the spot while calling less frequently and pecking gently at the back of her head. He then lowers

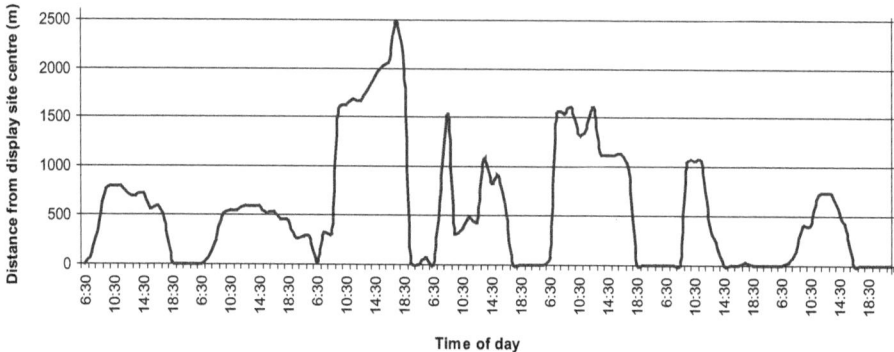

Figure 6.3: Movements of a satellite-tracked male bustard relative to the centre of its display site over several consecutive days during a peak display period in northern Australia show that its activity is focussed around its display site. It roosts overnight and then displays at the site in the morning, then moves away to forage and rest in adjacent cover before returning to the display site in the late afternoon.

himself on his tarsi, and with both birds extending their wings for balance, insemination occurs quickly.

Display bouts may last from a few minutes to two to three hours and mostly occur in the early morning and late afternoon (Figure 6.3). During peak display periods, males will often roost at their display sites overnight in order to begin displaying at first light. In warm climates, as the temperature rises during the morning, males will retreat to nearby shelter to rest during the heat of the day. They may return to their display sites later in the afternoon to recommence displaying. On cooler or overcast days displaying may last for longer periods and may occur throughout the day.

Display and nest site characteristics

Male bustards display to show off. Their large size, elaborate plumage and their low amplitude, but far-reaching call, all serve to make them stand out as much as possible. They therefore choose display sites that maximise their conspicuousness. Accordingly, sites are invariably in areas of low grass cover on open plains or in clearings in open shrub land and woodland habitats. If the opportunity exists, males will often display from an elevated podium such as a low rise, a furrow or a dam wall. Depending on the physical habitat, constraints of a site and the position of neighbouring males, the dimensions of the core display area may vary. On open plains, males with no competition from neighbours may display anywhere within a 100 to 200 metre radius of a central point. At sites with more wooded vegetation, core display sites may be more circumscribed and occur within the physical constraints of the surrounding habitat and be as small as a 20 x 20 metre patch.

Given their preference for open display areas, in more wooded habitats the size and configuration of leks may in large part be determined by suitable habitat conditions. In such areas where there are gradations in the favourability of sites, dominant males occupy the best sites while sub-ordinate males use sub-optimal sites.

In contrast to the open areas required for display, females prefer more sheltered areas for nesting and raising young. For this reason optimal breeding areas are those that balance the requirements of both sexes. The interfaces or ecotones between open grasslands and shrub or woodlands are therefore often the best areas for bustards because they include all the elements required for both sexes during the breeding season.

Timing of breeding

Given that the bustard's widespread distribution across Australia includes temperate, arid and tropical regions, the timing of breeding, which is dependent on productive environmental conditions, is variable. In southern temperate regions, breeding primarily occurs from July to December, although breeding in these regions is now a rare occurrence. In the more climatically variable arid interior of the continent, breeding is highly dependent on rainfall and may occur opportunistically at any time following sufficient rains. By comparison, in the northern wet-dry tropics breeding is highly seasonal and commences in the late dry season around September and may continue through until the end of the wet season in March. Invariably, breeding is timed to coincide with peak periods of food abundance, particularly of insects. However, the cues that stimulate breeding may vary between regions. In arid regions breeding is stimulated by rainfall; however, in the north males commence displaying up to two to three months before the onset of monsoonal rains. This suggests that they may in a sense 'anticipate' the upcoming period of high productivity and that there may be other cues, such as a rise in humidity and temperature, that stimulate the onset of display and breeding behaviour. Nevertheless, peak breeding periods in the north do not occur until the onset of monsoonal rains.

Nesting, eggs and raising young

True to form for a lekking species, following copulation the male has nothing else to do with the female or his offspring. The female alone assumes responsibility for incubation and raising young. She does not build a nest, but instead lays one or two, rarely three, eggs in a scrape on the ground. Nesting sites may be in bare open areas, beneath a shrub or tree, or among grass. There is some variation in egg colour but eggs are generally light olive-brown to olive-green and may be variously decorated with irregular spots, smears and longitudinal streaks. Incubation in captivity has been recorded as 24 days.

Bustard chicks are nidifugous (meaning they leave the nest soon after hatching) and precocial (meaning they are mobile and able to feed by themselves). They do, however, remain dependent on the female for several months and may sometimes remain with the parent up until the following breeding season. Following hatching, females lead young from the immediate vicinity of the nest to foraging areas. She collects food for the young or draws attention to edible items by pecking at them or dropping them in front of the chick. Fledging occurs after 30 to 35 days although at this stage the young are still only half-grown. Females may protect young by distraction displays whereby they pretend to be injured, or by chasing or shock displays aimed at intimidating or confusing intruders or predators.

Breeding success among Australian bustards is virtually unknown. In southern regions eggs and young may be taken by foxes. Other predators include raptors, ravens, dingoes, goannas and humans. Some reports suggest that females are easily disturbed off the nest by people, and this is likely to be more common in settled regions.

7

MOVEMENTS

The whole secret of life in arid regions is movement,
a readiness and a freedom to migrate.

F Debenham

The Australian continent's isolation, relatively limited land area and extremes of climatic and environmental variability are among the primary factors that drive the many and varied adaptations of its flora and fauna. Australia's birds are adapted to cope with the environment's variability and extreme conditions through a variety of strategies. Species or populations may exhibit boom-bust population fluctuations, switch diets to exploit available resources opportunistically, be able to tolerate lean periods, change activity rates, or move to more productive regions as required. While these strategies combine in different ways to determine an animal's ultimate response to its environment, for birds in particular, among the most important mechanisms for coping with change is their mobility. A notable outcome of the ability to fly is that birds can travel over vast distances quickly allowing them to make use of food and favourable habitat conditions over broad areas. Australia's birds exhibit a variety of movement strategies in response to changing environmental conditions. In contrast to the predominantly seasonal migratory movements of many birds in other regions, for example, the 'traditional' north-south migrants of Europe and North America, a significant proportion of Australian birds are characterised by nomadic, irruptive or partial movements, with movements across the landscape often occurring to exploit favourable conditions opportunistically. However, the details and drivers of these movements generally remain poorly understood.

Complex movement strategies

The Australian bustard is representative of a suite of species that undertake such complex movements in response to variable climatic conditions. The bustard has long been regarded as primarily dispersive or nomadic, undertaking irregular widespread movements over long distances. However, while the species certainly moves readily over the landscape, its widespread distribution, which spans regions that vary substantially in climatic and environmental conditions, means that it exhibits variable movements across its range, with nomadism representing only one type of movement strategy. Notably in regions influenced by productive and predictable seasonal conditions, bustards may be predominantly sedentary or their movements may tend towards regular migratory patterns as individuals move to exploit seasonally fluctuating resources. For example, in some recently cleared parts of eastern Queensland and in horticultural regions of northern Australia, bustards appear to be largely sedentary, staying put in areas all year round. By comparison, in temperate southern Australia some observers (though notably more so in the past when the abundance of bustards was greater in these regions than now) report seasonal patterns in abundance with numbers of bustards peaking over summer months, inferring seasonal influxes of bustards into these regions. Similarly, in the highly seasonal regions of the wet-dry tropics in northern Australia, there appear to be peaks in numbers in some regions coinciding with the onset of the wet season. Nevertheless, large-scale regular seasonal movements are more likely an exception than the norm.

In increasingly unpredictable environments, such as those that cover large parts of the bustard's inland range, individuals or populations may be increasingly mobile and nomadic, undertaking wide-ranging opportunistic movements in relation to shifting resources or habitat conditions that vary unpredictably or over longer climatic cycles. For example, in the Barkly Tableland region in the late dry season, bustards move between patches of highly productive areas following localised storms and rainfall. Similarly, bustards may move opportunistically in response to outbreaks of grasshoppers or mice plagues often into regions where they may not have been seen in for many years. Other dispersive, opportunistic movements may occur in response to deteriorating conditions, such as when bustards are displaced from regions affected by drought.

In effect then, the movement patterns of bustards span a gradient from sedentariness or regular migration in seasonally predictable regions to more idiosyncratic dispersive movements in increasingly climatically variable parts of the continent. Overall the movement strategies of bustards appear to be highly flexible and the boundaries defining movement types fluid, with birds exhibiting varying degrees of sedentariness, partial migration and nomadism according to climatic cycles and prevailing conditions, both locally and further afield.

Studying bustard movements

Until recently much of what we have been able to discern about the movements of the Australian bustard has been based on anecdotal observations and opportunistic records. However, such methods are inherently limited and they tell us little of the demographics of moving birds and the details of individual movements. More recently other methods have been employed to shed further light on the movements of bustards and other similarly mobile birds. Among these have been the use of systematic broad-scale surveys aimed at assessing changes in densities of bustards at specific locations over time and the use of radio and satellite telemetry to delineate local and broad-scale movements of individual birds.

Bird atlases and broad-scale surveys of rangeland users

National bird atlas and other coordinated bird surveys have greatly increased our general knowledge of Australia's birds and have demonstrated the utility of using the public in large-scale monitoring programs. Such schemes have facilitated broad-scale spatial and temporal analyses of Australian bird distributions and have propagated the development of specific analytical tools to describe large-scale movement patterns. One of the most notable recent efforts combined presence/

absence data from several databases to detect and describe broad-scale movement patterns for a large number of bird species in eastern Australia. The study provided valuable information regarding the shifting distribution patterns of birds over large areas over an annual cycle, and recognised a variety of movement strategies adopted by birds in the region. However, one of the major challenges for such techniques is the difficulty of acquiring enough data from remote regions. Presently such analyses are limited to species and regions for which there are adequate records, which largely excludes vast areas of the outback for which there are few records. Similarly, for species such as the Australian bustard that make variable or irregular movements analyses are limited because they require pooling of data into periods that are largely out of synchrony with temporal patterning of bustard movements. That is, movements may be responses to climatic variation over longer or less regular periods than seasonal or annual cycles.

So how do we collate adequate knowledge of bustard population dynamics and movements over a continental scale? Surveys of landholders and other rangeland users may provide a means of collecting information in remote areas to help detect patterns over broader spatial and temporal scales. Such surveys involving landholders and the public have been widely used to document distribution and population status of numerous taxa, including other bustard species elsewhere. In inland Australia, mail surveys of rangeland residents have been used to assess the distribution patterns, movements and population dynamics of the budgerigar *Melopsittacus undulates* over near continental scales. These proved highly effective in documenting the cyclical seasonal patterns of budgerigar movements in inland Australia. Similar recent efforts for the Australian bustard have produced comparable results. The most salient findings confirmed general predictions: that residency patterns of bustards vary widely across Australia. Seasonality of occurrence was more pronounced in regions characterised by predictable conditions. Interestingly, in more climatically unpredictable regions, there are also broad seasonal patterns to movements, although here they may be overlaid by more idiosyncratic movements in response to longer-term variation in rainfall and associated patterns of resource availability.

Satellite tracking of mobile fauna

Recent advances in satellite telemetry have provided biologists with unprecedented opportunities for examining the movement patterns of fauna over broad spatial and temporal scales. Satellite tracking studies have been used on a range of birds, particularly in the northern hemisphere, and have been instrumental in defining the large-scale migration movements of many highly mobile species. Tracking birds using satellites involves fitting a small transmitter unit to an individual which emits a signal that allows the geographical position of the animal to be determined by a satellite (or satellites) overhead. The location data are then sent to a receiving station back on the ground which distributes the information to the biologist. In effect, once the transmitter unit is deployed, the individual bird is tracked remotely with the biologist simply sitting at his or her computer and downloading the information via the internet.

The latest telemetry technologies are based on GPS which in theory allow for 24-hour-per-day position estimates and high levels of location accuracy (to within 10 metres of the true location 95% of the time). Transmitter units may also be powered by small solar panels that recharge internal batteries enabling tracking to be conducted continuously for several years. The spatial resolution of GPS data provides unprecedented levels of detail for examining space use and movements of wildlife, while the extended periods over which information can be gathered allow tracking studies to be conducted over several seasons. Such longer term tracking is particularly useful for species that move in relation to longer and/or less regular periods than seasonal or annual climatic cycles.

Satellite tracking may also be combined with spatial mapping technologies such as geographic information systems (GIS) and remote sensing, to provide powerful tools for relating movements to environmental conditions and habitat quality. In some cases movements can be related to events in near-real time. For example, the movements of bustards have been recently related to fires that are mapped by satellite (see below). In other cases, the distribution and movement patterns of highly mobile fauna may be related to rainfall or remotely derived indices of vegetation greenness or productivity over vast scales. Transmitter units may also be fitted with a range of options that measure specific aspects of an individual's activity or behaviour. Such studies, while still in their infancy in Australia, promise an exciting and revolutionary era in the way scientists study the movements of Australia's highly mobile birds.

Satellite tracking of bustards

The bustard is well suited to satellite tracking studies because of its large size (enabling long-term tracking using large transmitters – see Figure 7.1) and because of its preference for open spaces, which allows solar-powered

Figure 7.1: A male Australian bustard fitted with a solar-powered GPS satellite transmitter (arrow). Photo by Bruce Doran

tracking units to be readily recharged and minimises the potential impediment of signal transmissions due to overhead vegetation.

A recent satellite tracking study aimed to explore patterns of bustard movements in different regions that spanned a gradient from the highly predictable rainfall regions of monsoonal northern Australia to the highly variable climatic regions of the arid centre. All five individuals that were tracked in the northern regions were relatively sedentary making only local movements, with the longest movements spanning no more than 30 kilometres. Two of these individuals (both females) were tracked for periods of up to four years and averaged five location fixes per day. Both females were largely sedentary using relatively circumscribed areas but made short-scale movements in relation to season, local habitat conditions and fires.

Two bustards tracked in arid regions characterised by higher rainfall variability moved significantly further. One of these, a female fitted with a transmitter near Bedourie on the edge of the Simpson Desert in south-west Queensland, made extensive movements following a period of unusually high rainfall in the region (Colour Plate 8b). This area is among the most climatically variable regions on the continent with low and highly erratic rainfall resulting in patches of variable habitat quality. Soon after deployment of the tracking unit, and as the area began drying out, this female flew north in a series of steps over several days moving a maximum of 50 kilometres per day. Notably, these movements were made overnight with the female apparently resting during daylight hours. She eventually settled in an area of the Barkly Tableland, 500 kilometres north of where she had started. Towards the end of the wet season this individual subsequently moved south once more, retracing her original northward movements. Notably, her north-south movements followed the drainage lines of the Georgina River that form part of the Channel Country floodplains. These drainage channels represent important refuge areas as they are often more productive than surrounding areas and may therefore serve to guide or facilitate movements across the landscape.

Movements in relation to fire

Fires are a natural feature of the Australian landscape and play an important role in the functioning of ecosystems. Australia's flora and fauna are well-adapted to fire. Indeed, many bird species depend on fires for food and other resources and may readily track fires across the landscape. Bustards are attracted to fires and large numbers will often congregate in recently burnt areas to forage at fire fronts or in recently burnt areas for invertebrates and small vertebrates escaping or killed by fires. These areas are often exploited for several days following a fire until resources are depleted and the birds move on again (Figure 7.2).

The satellite tracking study of bustards discussed above also specifically investigated the response of individuals to local fires in the tropical savannas. Bustards moved directly to fires on several occasions with the longest direct linear movement to a fire of approximately 20 kilometres. Notably, however, not all fires were responded to, even when they occurred only within a kilometre or two of a tracked individual. This suggests that bustards may be either selective in their decisions to move or simply do not detect some fires. Detectability of a fire decreases as a function of distance and may depend on factors such as fire type, timing, size and wind direction. To detect fires bustards may use both visual and olfactory cues. Results from the satellite tracking study revealed that all movements directed towards fires were made when tracked individuals were downwind of a fire, suggesting that olfactory cues may be particularly important in detecting fires. In this respect these cues may be used especially in wooded habitats where smelling smoke may be easier than seeing a fire that is obscured by vegetation.

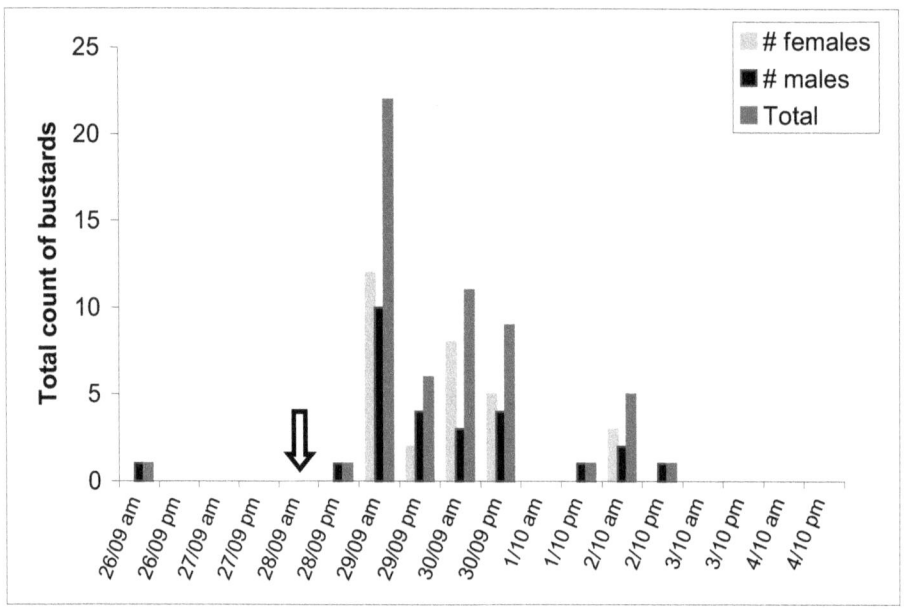

Figure 7.2: Response of bustards, measured as total number of males, females and all individuals, to a fire on a 4 km^2 plot in open woodland in northern Australia during the late dry season. Arrow indicates initiation of fire on the morning of 28 September.

Fires may also be more important to bustards at particular times of the year. In the highly seasonal savannas of northern Australia there is a general glut of food resources in the late wet and early dry season. As the dry season progresses, resource availability declines to the point that by the late dry season, resource availability is low, and the period represents a general time of hardship for many animals. Fires at this time may provide important opportunities for bustards because they provide valuable alternative foraging areas and easier access to foods. In contrast, foraging at fires in the early dry season may be less important because of the high abundance of food resources across the entire savanna landscape.

Irruptive movements

Bustards may make irruptive movements that are largely determined by climatic conditions. Irruptions among birds are generally triggered by high population levels as a consequence of prolonged or sequentially highly productive periods whereby numbers build up in an area. A subsequent reduction in food resources, such as may occur during droughts, then force *en masse* movements of large numbers of individuals into other areas. Among birds, irruptive movements mostly occur irregularly at intervals of a few years or decades, and may occur in apparently random directions with frequent underlying movements towards the coast from the inland.

Localised movements and home ranges

At the start of the breeding season in northern Australia, breeding female bustards begin to move extensively in their local area. For a short period they exhibit a spike in the daily distances they cover, which is often followed by a period of limited movement (Figure 7.3). There are two potential explanations for these fluctuations in movements. A characteristic of the northern savannas during the late dry/wet season transition is low food resource availability. Localised storms and patchy rainfall at this time result in patches of productive areas distributed across the landscape. Lower or patchy food availability may require individuals to range further or increase activity to satisfy nutritional requirements. However, in this case, increases in movement occur suddenly rather than gradually, even though a gradual increase may be expected given the steady decline of food resources as the dry season progresses. Also notable is that there is no commensurate increase in time spent foraging by bustards in the late dry season compared to periods of high food resource abundance earlier in the year.

A more plausible explanation for increases in female movements relates to the mating system of the bustard. The late dry season signifies the onset of the breeding season when male bustards begin displaying at exploded leks. Females range between these dispersed display sites and between leks to assess and choose potential mates. They therefore move over broader areas than they would otherwise. Once they have mated, they then settle down to nest, explaining why an increase in movements is followed by a lull in movements as females commence incubation and restrict movements to the nest and its immediate surroundings.

In comparison to the relatively extensive local movements made by females in the early breeding season, the movements and home ranges of mature males are considerably smaller. Males at this time spend limited time foraging, hence do not range in search of food, focusing their energies instead on attracting mates and on competing against rival males within the close vicinity of their display sites.

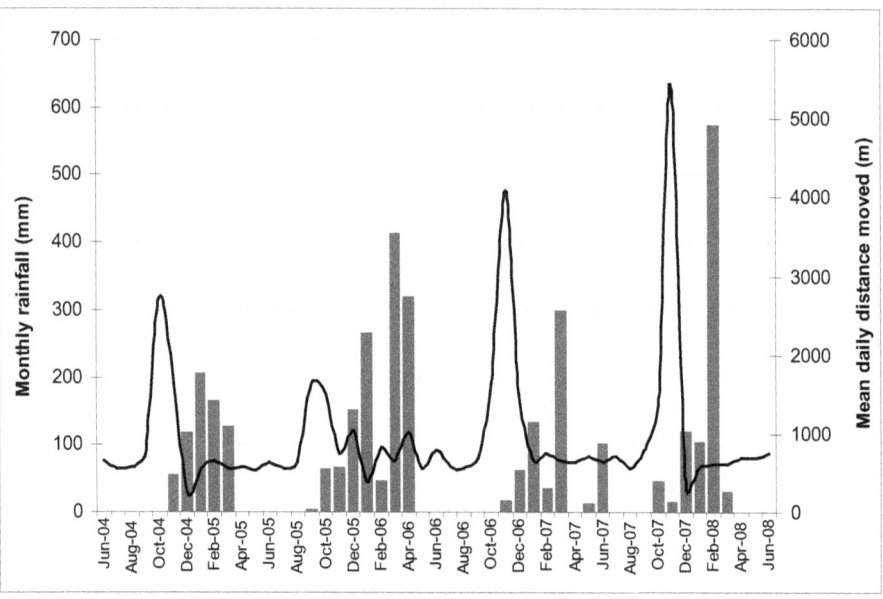

Figure 7.3: Mean daily distance moved per month (line) by a female bustard tracked by satellite in relation to rainfall (bars). The regular seasonal peaks in the late dry season coincide with the commencement of the breeding season and resource lows. The troughs following periods of extensive movements in most years relate to nesting events.

Post-breeding season movements

In the build up to and during the breeding season, bustards exhibit strong conspecific attraction (meaning they are attracted to their own kind) and aggregate at traditional lek sites. On a landscape scale these regions represent high densities of bustards. Following the breeding season in northern Australia, conditions are broadly productive across the entire savanna landscape as a result of the wet season. Bustards, uninhibited by the constraints of social and mating behaviour at this time, disperse over broader areas in order to take advantage of opportunities to exploit productive conditions with reduced competition.

It appears that although all elements of the population may move, it is primarily the younger, subordinate males that are most mobile following the breeding season in the north. In comparison, females with young and most larger males tend to be sedentary. An explanation for the difference between males relates to territoriality. Mature males are more territorial and remain near their display sites, whereas it is the younger males without established display territories that move. This appears consistent with the general hypothesis explaining movement in relation to territoriality whereby, among partially migrant populations, dominant individuals are expected to stay close to breeding territories, while subordinate individuals are more likely to move as a consequence of intra-specific competition. Following this period of post-breeding season dispersal, bustards then tend to move back towards traditional lekking areas in preparation for the following breeding season and the cycle starts anew. Similar post-breeding season movements, with a notable bias towards movements among males, have been reported among other bustard species including the great bustard, little bustard and the Bengal florican.

To walk or fly?

In the course of their daily routine, bustards generally prefer to walk rather than fly to get around. It has even been suggested that some individuals may go for days or even weeks without taking wing. Flight is more likely in the cooler morning or late afternoon and is generally avoided in the warmer parts of the day. The bustard's reluctance to fly when warm even extends to its strategies for predator avoidance. For example, they are more likely to take wing when approached in a vehicle in the cooler period of the early morning or late afternoon than in the heat of the day when they are more inclined to rely on camouflage to hide and stalk quietly away.

Take-off, particularly for large males, appears somewhat of a laborious affair. Still, windless conditions are especially challenging perhaps accounting in part for the bustard's preference to walk whenever possible. Once airborne, however, bustards are strong fliers and they tend to fly relatively low to the ground, within a few tens of metres above the surface. They have wide long wings with fingered wing tips and a broad wingspan that may reach up to 2.5 metres in large males. Their large size and slow wing beats makes for an impressive sight once in the air.

8

THREATS AND CONSERVATION

The imperious Bustard strides no more
Across the grassy waste;
The gallant Ruff deserts the shore
He trampled into paste;
The Oriole falls, a flaming sprite,
Before the unsparing gun;
Whilst thou by some diviner right
Dost wanton in the sun.

Arthur Benson 'The Sparrow'

The Australian bustard's decline from regions where it was formerly common is cause for concern and a sign of its susceptibility to changes wrought since European settlement. These declines have been variously attributed to hunting, pesticides, altered fire regimes, pastoralism, disturbance, habitat alteration and predation. These factors assume varying importance in different parts of the bustard's range and combine in different ways to affect local populations.

Reflecting the bustard's variable contemporary distribution across Australia, and partly owing to different classification criteria, the species' conservation status differs between jurisdictions. At present, it is not listed under the *Environment Protection and Biodiversity Conservation Act 1999*, but is considered 'Near Threatened' on a national scale according to the Action Plan for Australian Birds. More locally, the bustard is considered 'Critically Endangered' in Victoria, 'Endangered' in New South Wales and 'Vulnerable' in South Australia and the Northern Territory.

> *It may be possible – and, indeed it is most likely – that … no longer does the noble Bustard stalk over the flats of the … Hunter … and if this be so, surely the Australians should at once bestir themselves to render protection of these and many other native birds: otherwise very many of them … will soon become extinct.*
>
> *John Gould 1865*

Hunting

It was recognised early following European settlement that bustards were in decline and in need of protection. Following the arrival of settlers, the first and most profound impact on bustards was through excessive hunting. The renown naturalist, John Gould, recognised the trajectory of some native bird populations in Australia and was well positioned to comment. He had himself borne witness to the decline of the great bustard in his native England which officially bred for the last time in Britain in 1832 – just a few years prior to Gould's departure on his pioneering trip to Australia. The first formal attempt to protect the Australian bustard was a schedule protecting game within the New South Wales Act that came into force in April 1866. The equivalent Victorian Act was introduced a year later. These legislative initiatives had decidedly pragmatic and anthropocentric motivations because they were introduced to protect declining game species that were useful to local settlers, including imported species that had been deliberately introduced for hunting. Under the Act in Victoria, hunting was prohibited during the breeding season from the beginning of August to the end of December each year on pain of a £2 penalty for taking birds outside the allowable season. By 1902, shooting bustards was prohibited altogether in Victoria, and the species has been formally protected since 1940. However, despite protection, the persecution of the bustard continued, all too often with an indifferent, sometimes openly defiant attitude on the part of the public.

> *A very general attitude towards the law was expressed by a normally very law-abiding citizen with whom I discussed 'Wild Turkey'. Said he – 'It's the open season whenever I see one'.*
>
> *EH Sedgewick 1954*

But it was not only 'normally very law-abiding citizens' that were remiss of the law. Local law enforcers were sometimes complicit, if not directly involved in shooting the protected species. Several accounts in the 1930s recount stories of local police hunting bustards. In one case, inmates at a local jail were regularly fed 'turkey dinners' after the local constables went out shooting on the weekends.

One reason for the susceptibility of the bustard to hunting is the ease with which they can be approached in a vehicle. It is often possible to get within a few metres of an individual, particularly in the heat of the day when they are more likely to opt to stay still and hide than to take flight.

> *… although the bustard is usually very shy and wary in the presence of man on foot or horse-back … it never seems to realise the danger from a car, but stands, with head in the air, to be shot, a mark not to be missed at close range.*
>
> *FL Berney 1936*

Inevitably, when the opportunity arose, the ease of hunting and local abundance of bustards in some parts led to excess and over-exploitation. FL Berney writes on another occasion of a car that came into Longreach, central Queensland, that was 'so covered with plains turkeys as to have its outline obscured'.

Although hunting by the general population is prohibited, and attitudes have changed in recent decades, illegal hunting still occurs. Hunting is also permitted and continues to varying degrees by Aboriginal people in

Figure 8.1: An old NSW National Parks and Wildlife Service poster highlighting that hunting of bustards is illegal in the state.

central and northern Australia. Breakdown over controls of traditional hunting, coupled with potentially greater hunting efficiency facilitated by motor vehicles and guns, means that hunting may continue to have significant, localised impacts on bustard populations.

Apart from the ease with which bustards can be harvested with contemporary technology, there are inherent biological reasons for the susceptibility of bustard populations to over-exploitation. First among these is their low reproductive output. As a long-lived and late maturing species with few young at a time, individuals are not quickly replaced. Furthermore, in the more arid parts of their range, breeding is largely opportunistic and dependent on favourable environmental conditions. Since these conditions are largely dependent on rainfall, which may be highly variable, breeding may not occur every year in these parts of the country.

The bustard may be particularly susceptible to biased harvesting or over-exploitation on account of its lek mating system. During the breeding season males develop spectacular plumage and display noisily in open or raised areas, making them particularly conspicuous. Females, by comparison, are smaller, more drab in appearance and shy, relying on their camouflage to protect their nest and young. The largest males, which are responsible for most of the breeding success in lekking populations, are therefore the most likely to be harvested because they are more visible and their great size makes them prized game. Additionally, the tendency for bustards to exhibit strong fidelity to lekking grounds means that if these sites are known, they are at risk of over-harvesting. In effect, leks may then act as 'sinks' because as individuals are removed new individuals move in to occupy vacated display sites. However, studies of other bustard species suggest that once leks become extinct through habitat alteration or over-exploitation, they are not re-established. Thus, uncontrolled or illegal hunting may not only lead to direct reduction in overall bustard numbers but to a significant loss of prime males which disrupts the social structure that forms the basis of the lek mating system. In the longer term, the genetic viability of populations is potentially comprised by the selective removal of the largest and fittest of males. Indeed, the effects of excessive or biased harvesting among such lekking species may have disproportionally adverse consequences compared to species that have less specialised mating systems.

Introduced predators

Compounding the effects of hunting following settlement were the effects of animals that the settlers brought with them. The bustard is one of a large suite of ground-dwelling and nesting Australian birds that have been much reduced by introduced predators, particularly the introduced European red fox *Vulpes vulpes*. The areas where bustards have declined most correspond closely to the distribution of foxes in Australia with very few recent breeding records from areas where foxes occur. However, since the northward expansion of foxes is limited by climatic conditions, populations in the northern parts of the continent are not as affected by foxes. As ground-dwelling birds, it is the eggs, young and females at nests that are most susceptible to predation. The effects of feral cats and dingoes are not frequently cited as a threat for bustards but it is likely that they are also susceptible to these predators to some extent.

Pesticides

Bustards are susceptible to poisons through direct ingestion of poisoned baits or by the consumption of poisoned prey such as grasshoppers and mice. Hundreds, perhaps thousands, of bustards were reportedly killed by grain treated with arsenic, phosphorous and strychnine laid out over many thousands of kilometres to eradicate rabbits at the turn of the last century. Such poisoning is cited as having caused local extinctions in some districts.

> In 1882 when phosphorized oats were scattered in the Riverina, bustards died in hundreds. They were found dead on runs where no poison had been laid, and they had flown many miles after eating the grain.
>
> E Rolls 1969

The mobile nature of bustards, and their readiness to track outbreaks of favoured food resources across the landscape, means they are susceptible to pesticides used to control mice plagues and grasshoppers. Their tendency to gorge on such foods at these times means that their levels of exposure to toxins may be high. Furthermore, outbreaks of such prey represent 'boom' times to bustards and other birds that stimulate breeding. However, given that young are likely to be more susceptible to poisoned foods, and some pesticides are known to alter a bird's ability to metabolise calcium leading to thin eggshells, breeding events in such areas may be compromised.

Habitat alteration

Localised land clearing often benefits bustards by opening up habitats that are otherwise less preferred or avoided such as thick woodlands and forests. However, despite the initial, positive response of bustards to land clearing in some places, intensification of agriculture and land use following clearing has contributed to the decline of bustards in other regions where they were formerly common. Intensification may take various forms including the construction of roads and other infrastructure, use of pesticides, increased human traffic (including hunting pressure), and other factors associated with intensive farming practices. Furthermore, extensive, broad-scale clearing and cropping with single homogenous crops also reduces the diversity of habitats required by bustards. In some regions, harvesting of crops using heavy machinery also coincides with the breeding season of bustards and has the potential to destroy nests.

In direct contrast to the clearing and opening up of habitats, in some regions of the north the habitats of bustards are affected by the thickening of open areas due to 'woody weed' infestation. Thickening of vegetation in open woodland habitats has been recognised as a threatening process for several bird species. Its occurrence and intensity varies between habitats and appears particularly pronounced on the alluvial plains favoured by bustards. In the Victoria River District of the Northern Territory, for example, in just a few decades once broad open plains have become dense woodlands (Figure 8.2). Infestations have been primarily attributed to overgrazing by cattle which has led to a transformation of much of the herbaceous vegetation to a largely non-flammable state. This has resulted in significantly fewer effective fires, which along with an increase in monsoonal rainfall over the last few decades, appears to have resulted in higher rates of 'woody weed' establishment. As a consequence, large areas of formerly open plains and sparse open woodland are increasingly wooded. Such areas are less suitable as display grounds and in time may reach a threshold of shrub and tree density that bustards avoid altogether.

Altered fire regimes

Fire is a dominant, pervasive feature of the Australian environment that has been an important factor in shaping the evolution of Australia's fauna and flora. Species are generally dependent on particular types of fire regimes that may vary in fire frequency, extent of area burnt, intensity, seasonality, etc. Recent declines in the distribution and population sizes of several bird and mammal species throughout Australia have been at least in part linked to changes in fire regimes following European settlement.

The dominant outcome of traditional Aboriginal fire regimes was a fine-scaled mosaic of fire regimes whereby relatively small areas were burnt at different intervals and frequencies. In the wet-dry tropics of northern Australia, fires tended to be lit in the early dry season in order to clear areas of grassy undergrowth following the wet season and to facilitate new growth to attract game. Following the cessation of traditional Aboriginal fire regimes, fire management changed from primarily fine-

Figure 8.2: Progressive 'woody weed' infestation (primarily by native *Terminalia* spp.) of open plains at Kidman Springs in the Victoria River District, Northern Territory; 1968 (top), 1979 (middle) and 1998 (bottom). Photos by Brian Hill (top and middle) and by Rod Dyer (bottom)

scaled, early dry season, 'cool' fires, to large-scaled, late dry season, hot fires. Such a change is likely to have had a profound impact on the biota. Late dry season fires act as double-edged swords for bustards. Bustards forage at fire fronts or in recently burnt areas to exploit food resources exposed or killed by fires. Late season fires provide greater accessibility to foods at a time of low food availability across the landscape. Bustards therefore readily track fire events over local areas. However, late season fires also coincide with the nesting period for bustards therefore have the potential to destroy nests. In the longer term, changed fire regimes may affect bustards and other species by altering food resources, floristics, nutrient availability and habitat structure.

Other threats

The bustard's preference for open spaces often means birds are found along road corridors, particularly in more wooded regions. As a large bird with a slow, cumbersone take-off they are susceptible to vehicle strike, and the sight of scattered feathers following collisions are not uncommon in areas where they are often found.

Powerlines are a recognised problem for several bustard species. However, while the potential exists for collisions in some parts of the Australian bustard's range, powerline infrastructure in remote areas is generally scarce. Nevertheless, as a large bird with a broad wing span that tends to fly at a low altitude, strikes may occassionally occur.

Conservation efforts

The bustard's widespread distribution across Australia, its variable movement strategies and mating systems and the different threats that it faces mean that there are no simple, homogenous prescriptions for its protection. Conservation strategies therefore need to consider the particular issues facing bustards in specific areas, and attempt where possible, to take a holistic approach in tackling them.

Captive breeding

Re-introduction of bustards into areas from where they have disappeared has been proposed for several different bustard species. To this end, captive breeding programs have been initiated in several countries with the aim of breeding sufficient numbers of individuals to release into the wild. These programs have met with mixed success and have generally been regarded as part of holistic conservation efforts rather than an end in themselves. Invariably such programs are expensive, labour-intensive and require significant expertise.

In 1959 the Victorian Government established a wildlife research station known as 'Serendip Sanctuary' outside Melbourne with the objective of studying and rehabilitating species that had become rare or extinct in Victoria. One of the sanctuary's aims was to establish a captive breeding program for the Australian bustard with the hope of re-introducing individuals back into the wild in the state. In the mid-1960s the program began with a dozen wild caught birds from the Northern Territory and western South Australia. An elaborate enclosure was designed so that females could access males that were each housed separately, but that prevented males from direct contact with each other. The idea was to mimic, as far as was possible in captivity, the natural lek mating system of bustards by offering females the opportunity of mating with the male of their choice. Significant initial successes were achieved, with the population reaching a peak of 68 by 1989. However, following the departure of key personnel and expertise from the sanctuary, and compounded by a series of attacks by predators and natural losses over the following years, the birds ceased breeding and numbers dwindled to only 10 by the year 2000. Recent concerted efforts to revive the success of the early years have been promising, with breeding success again achieved in 2004.

Although there is some hope that a viable population in captivity can be maintained, it is doubtful that the original aim of re-introduction into the wild can be achieved, at least not in the short to medium term. Among the major issues are ensuring that captive bred individuals are conditioned to survive in the wild and that there is sufficient genetic diversity in the birds that are bred to be released. Most importantly, before any re-introductions can be contemplated, the original threats that led to the bustard's decline in the first place need to be adequately addressed to ensure that released bustards do not meet the same fate as the original southern populations endured.

The future?

The poorly understood, dispersive movement strategies that bustards employ across much of their range create special problems for their protection. Indeed, the protection of dispersive fauna in general is regarded as one of the greatest challenges for conservation biology because conventional conservation practices and representative reservation cannot adequately cater for such species. That is, because of their ability to move and track favourable environmental conditions over large areas, when conditions in an area or reserve deteriorate because of natural climatic cycles or events, mobile animals such as bustards simply move to other areas. They may then not be afforded the same protection as they may be within managed areas.

A fundamental problem is understanding how and why populations change and how the complex movement patterns of bustards and similar species are employed. As discussed in the previous chapter, one method for acquiring adequate knowledge of bustard population dynamics and movements may be through the broad-scale participation of landholders. Selected residents across remote regions may help monitor numbers of dispersive species on their properties, and combined with existing atlas survey techniques and scientific surveys, may provide an effective and unusually spatially representative and cost-effective monitoring system. Such initiatives may be combined with emerging technologies, such as satellite telemetry and spatial information systems, that allow tracking individuals and relating their movements to environmental conditions in near-real time.

If the locations and preferred habitats of dispersive species such as the bustard are known or can be predicted at particular times, then such sites can be prioritised for protection at critical times. Such knowledge would effectively facilitate predictive or pre-emptive conservation planning (e.g. by developing mobile or shifting conservation zones in time and space), thereby overcoming the limitations of current static reserve design and conservation strategies. For example, refugia required during drought, or breeding habitats and lekking areas used by bustards, could be protected at key times by controlling introduced predators, managing for appropriate fire regimes, minimising stocking rates for livestock, or implementing moratoria on hunting. Current strategies for the conservation of highly mobile fauna are largely inadequate, highlighting the need for lateral and creative approaches for their protection. The bustard, a lordly icon of the Australian outback, deserves nothing less. It sustained many early Australians, both Indigenous and colonising, and now it is our turn to ensure its survival.

BIBLIOGRAPHY

Albrecht G & Albrecht J (1992) The Goulds in the Hunter Region of N.S.W. 1839–1840. In *Naturae No 2*, pp. 24–25. Centre for Bibliographical and Textual Studies, Monash University.

Allan DG (1994) The abundance and movements of ludwigs bustard *Neotis ludwigii*. *Ostrich* **65**, 95–105.

Alonso JA, Martin CA, Alonso JC, Morales JM & Lane SJ (2001) Seasonal movements of male great bustards in central Spain. *Journal of Field Ornithology* **72**, 504–511.

Alonso JC & Alonso JA (1992) Male-biased dispersal in the great bustard *Otis tarda*. *Ornis Scandinavica* **23**, 81–88.

Alonso JC, Martin E, Alonso JA & Morales MB (1998) Proximate and ultimate causes of natal dispersal in the great bustard *Otis tarda*. *Behavioural Ecology* **9**, 243–252.

Alonso JC, Morales MB & Alonso JA (2000) Partial migration, lek fidelity and nesting area fidelity in female great bustards *Otis tarda*. *Condor* **102**, 127–136.

Alonso JC, Palacin C & Martin CA (2003) Status and recent trends of the great bustard (*Otis tarda*) population in the Iberian peninsula. *Biological Conservation* **110**, 185–195.

Anon. (2000) *Ngarrindjeri Dreaming Stories*. South Australian Department of Education, Training and Employment, Adelaide.

Appayya MK (1982) Breeding of bustards: an observation in Australia. *Journal of the Bombay Natural History Society* **79**, 195–197.

Badman FJ (1979) Birds of the southern and western Lake Eyre drainage. *South Australian Ornithologist* **28**, 29–55.

Barker RD & Vestjens WJM (1989) *The Food of Australian Birds. I. Non-passerines*. CSIRO Division of Wildlife and Ecology, Lyneham.

Barrett G, Silcocks A, Barry S, Cunningham RB & Poulter R (2003) *The New Atlas of Australian Birds*. Royal Australasian Ornithologists Union, Hawthorn East.

Bennett DH (1983) Some aspects of Aboriginal and non-Aboriginal notions of responsibility to non-human animals. *Journal of the Australian Institute of Aboriginal Studies* **2**, 19–24.

Benson AC (1895) *Selected Poems*. John Lane, London.

Berney FL (1907) Field notes on the birds of the Richmond District, north Queensland. *Emu* **6**, 106–115.

Berney FL (1936) The bustard in Queensland. *Emu* **36**, 4–9.

Blakers M, Davies SJJF & Reilly PN (1984) *The Atlas of Australian Birds.* RAOU and Melbourne University Press, Melbourne.

Boehm EF (1947) The Australian Bustard: with special reference to its past and present status in South Australia. *South Australian Ornithologist* **18**, 37–40.

Booth T (2006) The reminiscences of Tom Booth at 'Corrong' Station and elsewhere in the Hay District. *Hay Historical Society Website Newsletter IV*, <http://users.tpg.com.au/hayhist/NewsletterFour.html>.

Bowman DMJS (1998) Tansley Review No. 101. The impact of aboriginal landscape burning on the Australian biota. *New Phytologist* **140**, 385–410.

Brady CJ (2008) Male-male conflict and breeding of the Australian Bustard *Ardeotis australis* in rehabilitated mine land in Arnhem Land, Northern Territory. *Australian Field Ornithology* **25**, 203–206.

Branch H, *Old Adaminaby and Lake Eucumbene, including relics and moveable objects.* NSW Heritage Branch, Sydney, <http://www.heritage.nsw.gov.au/07_subnav_02_2.cfm?itemid=5060670>.

Bravery JA (1970) The birds of the Atherton Shire, Queensland. *Emu* **70**, 49–63.

Broders O, Osborne T & Wink M (2003) A mtDNA phylogeny of bustards (family Otididae) based on nucleotide sequences of the cytochrome b-gene. *Journal of Ornithology* **144**, 176–185.

Brooker MG, Ridpath MG, Estbergs JA, Bywater J, Hart DS & Jones MS (1979) Bird observations on the North-western Nullarbor Plain and neighbouring regions, 1967–1978. *Emu* **79**, 176–190.

Brown AG (1950) The birds of Turkeith, Victoria. *Emu* **50**, 105–113.

Campbell AG (1902) Birds of north-eastern Victoria. *Emu* **2**, 9–18.

Carranza J, Hidalgo SJ & Ena V (1989) Mating system flexibility in the great bustard: a comparative study. *Bird Study* **36**, 192–198.

Collar NJ (1996) Family Otididae (Bustards). In *Handbook of the Birds of the World. Volume 3: Hoatzin to Auks.* (Eds J del Hoyo, A Elliott and J Sargatal), pp. 240–273. Lynx Edicions, Barcelona.

Collins J, Klomp NI & Birckhead J (1996) Aboriginal use of wildlife: past, present and future. In *Sustainable Use of Wildlife by Aboriginal Peoples and Torres Strait Islanders.* (Eds M Bomford and J Caughley) pp. 14–36. Australian Government Publishing Service, Canberra.

Combreau O, Gelinaud G & Smith TR (2000) Home range and movements of houbara bustards introduced in the Najd Pediplain in Saudi Arabia. *Journal of Arid Environments* **44**, 229–240.

Combreau O, Launau F, Bowardi MA & Gubin B (1999) Outward migration of houbara bustards from two breeding areas in Kazakhstan. *Condor* **101**, 159–164.

Cunningham PM (1827) *Two Years in New South Wales: A Series of Letters.* Henry Colburn, London.

Debenham F (1954) The geography of deserts. In: *Biology of Deserts.* (Ed. JL Cloudsley-Thompson) pp. 1–12. Institute of Biology, London.

Dickison DJ (1932) History and early records of ornithology in Victoria. *Emu* **31**, 175–196.

Downes MC (1975) *A Bibliography of Bustards.* Department of Agriculture, PNG.

Downes MC (1982a) *The Australian Bustard in the Barkly region of the Northern Territory.* Report to the Conservation Commission of the Northern Territory, Darwin.

Downes MC (1982b) *A management programme for the Bustard in the Northern Territory.* Report to the Conservation Commission of the Northern Territory, Darwin.

Downes MC (1982c) Re-establishment of the Bustard in Victoria. In: *Wildlife Management in the '80s.* (Ed. T Riney). Proceedings of the Conference for the Field and Game Federation of Australia and Graduate School of Environmental Science, Monash University.

Downes MC (1984) *The Bustard in South Australia.* National Parks and Wildlife Service, Adelaide.

Downes MC & Speedie C (1982) *Classification of bustard habitat in the Northern Territory.* Report to the Conservation Commission of the Northern Territory, Darwin.

Fitzherbert JC & Baker-Gabb DJ (1988) Australasian grasslands and their threatened avifauna. In *Ecology and Conservation of Grassland Birds.* (Ed. PD Goriup) pp. 227–250. International Council for Bird Preservation, Cambridge.

Fitzherbert K (1978) Observations on breeding and display in a colony of captive Australian Bustards (*Ardeotis australis*). B.Sc.(Hons) thesis. Monash University, Clayton.

Fitzherbert K (1981) Seasonal weight changes and display in captive Australian bustards (*Ardeotis australis*). In *Bustards in Decline.* (Eds PD Goriup and H Vardhan) pp. 210–225. Tourism and Wildlife Society of India, Jaipur.

Flinders M (1814) *A Voyage to Terra Australis.* G & W Nicol, London.

Franklin DC (1999) Evidence of disarray amongst granivorous bird assemblages in the savannas of northern Australia, a region of sparse human settlement. *Biological Conservation* **90**, 53–68.

Garnett ST & Crowley GM (2000) *The Action Plan for Australian Birds*. Environment Australia, Melbourne.

Gaucher P (1995) Breeding biology of the houbara Bustard *Chlamydotis undulata undulata* in Algeria. *Alauda* **63**, 291–298.

Goriup PD (1988) *Ecology and Conservation of Grassland Birds*. International Council for Bird Preservation, Cambridge.

Goriup PD (1994) Little Bustard *Tetrax tetrax*. In *Birds in Europe: Their Conservation Status* (GM Tucker and MF Heath) pp. 236–237. Birdlife International, Cambridge.

Goriup PD & Vardhan H (1980) *Bustards in decline*. Tourism and Wildlife Society of India, Jaipur.

Gould J (1865) *Handbook of the Birds of Australia, Vol. 1*. 1972 facsimile edition. Landsdowne, Melbourne.

Grice D, Caughley G & Short J (1986) Density and distribution of the Australian Bustard *Ardeotis australis*. *Biological Conservation* **35**, 259–267.

Griffioen P & Clarke MF (2002) Large-scale bird-movement patterns evident in eastern Australian atlas data. *Emu* **102**, 99–125.

Hallager SL (1994) Drinking methods in two species of bustards. *Wilson Bulletin* **106**, 763–764.

Hingrat Y, Saint Jalme M, Ysnel F, Lacroix F, Seabury J & Rautureau P (2004) Relationships between home-range size, sex and season with reference to the mating system of the Houbara Bustard *Chlamydotis undulata undulata*. *Ibis* **146**, 314–322.

Hingrat Y, Saint Jalme M, Ysnel F, Le Nuz E & Lacroix F (2007) Habitat use and mating system of the houbara bustard (*Chlamydotis undulata undulata*) in a semi-desertic area of North Africa: implication for conservation. *Journal of Ornithology* **148**, 39–52.

Hingrat Y, Ysnel F, Saint Jalme M, Le Cuziat J, Béranger P-M & Lacroix F (2007) Assessing habitat and resource availability for an endangered desert bird species in eastern Morocco: the Houbara Bustard. *Biodiversity and Conservation* **16**, 597–620.

Isakov YA (1974) Present distribution and population status of the great bustard, *Otis tarda* Linnaeus. *Journal of the Bombay Natural History Society* **71**, 433–444.

Jiguet F (2002) Arthropods in diet of Little Bustards *Tetrax tetrax* during the breeding season in western France. *Bird Study* **49**, 105–109.

Jiguet F, Arroyo B & Bretagnolle V (2000) Lek mating systems: A case study in the Little Bustard *Tetrax tetrax*. *Behavioural Processes* **51**, 63–82.

Jiguet F & Bretagnolle V (2001) Courtship behaviour in a lekking species: individual variations and settlement tactics in male little bustard. *Behavioural Processes* **55**, 107–118.

Johnsgard PA (1991) *Bustards, Hemipodes, and Sandgrouse: Birds of Dry Places*. Oxford University Press, Oxford.

Judas J, Combreau O, Lawrence M, Saleh M, Launay F & Xingi G (2006) Migration and range use of Asian Houbara Bustard *Chlamydotis macqueenii* breeding in the Gobi Desert, China, revealed by satellite tracking. *Ibis* **148**, 343–351.

Lane SJ, Alonso CJ & Martin CA (2001) Habitat preferences of great bustard *Otis tarda* flocks in the arable steppes of central Spain: are potentially suitable areas occupied? *Journal of Applied Ecology* **38**, 193–203.

Lane SJ & Alonso JC (2001) Status and extinction probabilities of great bustard (*Otis tarda*) leks in Andalucia, southern Spain. *Biodiversity and Conservation* **10**, 893–910.

Lane SJ, Alonso JC, Alonso JA & Naveso MA (1999) Seasonal changes in diet and diet selection of great bustards (*Otis. tarda*) in north-west Spain. *Journal of Zoology* **247**, 201–214.

Launay F, Combreau O & Al Bowardi M (1999) Annual migration of Houbara Bustard *Chlamydotis undulata macqueenii* from the United Arab Emirates. *Bird Conservation International* **9**, 155–161.

Launay F, Combreau O, Aspinall SJ, Loughland RA, Gubin B & Karpov F (1999) Trapping of breeding houbara bustard (*Chlamydotis undulata*). *Wildlife Society Bulletin* **27**, 603–608.

Launay F, Roshier D, Loughland R & Aspinall SJ (1997) Habitat use by houbara bustard (*Chlamydotis undulata macqueenii*) in arid shrubland in the United Arab Emirates. *Journal of Arid Environments* **35**, 111–121.

Lavee D (1985) The influence of grazing and intensive cultivation on the population size of the houbara bustard in the northern Negev in Israel. *Bustard Studies* **3**, 103–107.

Lichtenberg EM & Hallager SL (2006) A description of commonly observed behaviors for the kori bustard (*Ardeotis kori*). *Journal of Ethology* **26**, 17–34.

Lloyd P, Little RM, Crowe TM & Simmons RE (2001) Rainfall and food availability as factors influencing the migration and breeding activity of Namaqua Sandgrouse, *Pterocles namaqua*. *Ostrich* **72**, 50–62.

Mac Nally R, Ellis M & Barrett G (2004) Avian biodiversity monitoring in Australian rangelands. *Austral Ecology* **29**, 93–99.

Marchant S & Higgins PJ (1993) *Handbook of Australian, New Zealand and Antarctic Birds. Vol.2. Raptors to Lapwings.* Oxford University Press, Melbourne.

Marshall AJ (1932) The position of the Australian Bustard today. *Emu* **32**, 84–86.

Martinez C (1991) Patterns of distribution and habitat selection of a great bustard (*Otis tarda*) population in north-western Spain. *Ardeola* **38**, 137–147.

Martinez C (1992) Variation in flock size and flock type of great bustard (*Otis tarda*) according to habitat. *Miscellania Zoologica* **16**, 161–170.

Martinez C (1994) Habitat selection by the little bustard *Tetrax tetrax* in cultivated areas of Central Spain. *Biological Conservation* **67**, 125–128.

Martinez C (2000) Daily activity patterns of the Great Bustard *Otis tarda*. *Ardeola* **47**, 57–68.

Mattingley AHE (1929) The love-display of the Australian Bustard. *Emu* **28**, 198–199.

McMillan RP (1950) Food of the Bustard. *The Western Australian Naturalist* **2**, 10–11.

Morales MB (1999) Ecología reproductiva y movimientos estacionales en la avutarda (*Otis tarda*). Tesis doctoral. Universidad Complutense, Madrid.

Morales MB, Alonso JC & Alonso J (2002) Annual productivity and individual female reproductive success in a Great Bustard *Otis tarda* population. *Ibis* **144**, 293–300.

Morales MB, Alonso JC, Alonso JA & Martin E (2000) Migration patterns in male great bustards. *Auk* **117**, 493–498.

Morales MB, Alonso JC, Martin C, Martin E & Alonso J (2003) Male sexual display and attractiveness in the great bustard *Otis tarda*: the role of body condition. *Journal of Ethology* **21**, 51–56.

Morales MB, Jiguet F & Arroyo B (2001) Exploded leks: what bustards can teach us. *Ardeola* **48**, 85–98.

Moreira F (2004) Distribution patterns and conservation status of four bustard species (Family Otididae) in a montane grassland of South Africa. *Biological Conservation* **118,** 91–100.

Morgado R & Moreira F (2000) Seasonal population dynamics, nest site selection, sex-ratio and clutch size of the Great Bustard *Otis tarda* in two adjacent lekking areas. *Ardeola* **47**, 237–246.

Morton SR, Short J & Barker RD (2001) *Refugia for biological diversity in arid and semi-arid Australia*. Environment Australia, Canberra.

Mountford CP (1949) Gesture language of the Warlpiri tribe, central Australia. *Transactions of the Royal Society of South Australia* **73**, 100–101.

Mwangi EM & Karanja WK (1989) Home range, group size and sex composition in white bellied and kori bustards. *Bustard Studies* **4**, 114–122.

Nadeem MS, Ali F & Akhtar MS (2004) Diet of the Houbara Bustard *Chlamydotis undulata* in Punjab, Pakistan. *Forktail* **20**, 91–93.

Osborne PE, Al Bowardi M & Bailey TA (1997) Migration of the houbara bustard *Chlamydotis undulata* from Abu Dhabi to Turkmenistan: the first results from satellite tracking studies. *Ibis* **139**, 192–196.

Osborne PE, Alonso JC & Bryant RG (2001) Modelling landscape-scale habitat use using GIS and remote sensing: a case study with great bustards. *Journal of Applied Ecology* **38**, 458–471.

Osborne PE, Bryant RG & Alonso JC (1998) Application of a time-series of AVHRR data to mapping the habitat of great bustards in Spain. *Proceedings of the 24th Annual Conference of the Remote Sensing Society*, pp. 136–142.

Osborne PE, Launay F & Gliddon D (1997) Wintering habitat use by houbara bustards *Chlamydotis undulata* in Abu Dhabi and implications for management. *Biological Conservation* **81**, 51–56.

Osborne T. & Osborne L. (1998) *Ecology of the Kori bustard in Namibia*. First annual report for the Ministry of Environment and Tourism Permit Office, Windhoek.

Osborne T & Osborne L (2001) *Ecology of the Kori bustard in Namibia*. Annual report of the Ministry of Environment and Tourism Permit Office, Windhoek, Namibia.

Palacios M, Garzon J & Castroviejo J (1975) La gestión en la Reserva de la avutarda (*Otis tarda*) en España, especialmente en primavera. *Ardeola* **21**, 347–406.

Ponjoan A, Botab G, García De La Morenac EL, Morales M, Wolffe A, Marco I & Manosa S (2008) Adverse effects of capture and handling Little Bustard. *Journal of Wildlife Management* **72**, 315–319.

Rolls E (1969) *They All Ran Wild: The Story of Pests on the Land in Australia*. Angus and Robinson, Sydney.

Salamolard M & Moreau C (1999) Habitat selection by Little Bustard *Tetrax tetrax* in a cultivated area of France. *Bird Study* **46**, 25–33.

Seddon PJ, Launay F, van Heezik Y & Al Bowardi M (1999) Methods for live trapping Houbara Bustards. *Journal of Field Ornithology* **70**, 169–181.

Seddon PJ & Van Heezik Y (1996) Seasonal changes in Houbara bustard *Chlamydotis undulata macqueenii* numbers in Harrat al Harrah, Saudi Arabia: Implications for managing of remnant population. *Biological Conservation* **75**, 139–146.

Sedgewick EH (1954) The Australian bustard in Western Australia. *Emu* **54**, 181–186.

Serventy DL & Whittell HM (1976) *Birds of Western Australia*. University of Western Australia, Perth.

Silva JP, Faria N & Catry T (2007) Summer habitat selection and abundance of the threatened little bustard in Iberian agricultural landscapes. *Biological Conservation* **139**, 186–194.

Silva JP, Pinto M & Palmeirim JM (2004) Managing landscapes for the little bustard *Tetrax tetrax*: lessons from the study of winter habitat selection. *Biological Conservation* **117**, 521–528.

Smith PJ, Smith JE, Pressey RL & Whish GL (1995) Birds of particular conservation concern in the western division of New South Wales: distributions, habitats and threats. In *Occasional Paper 20*, p. 91. NSW National Parks and Wildlife Service, Hurstville.

Suarez-Seoane S, Osborne PE & Alonso JC (2002) Large-scale habitat selection by agricultural steppe birds in Spain: identifying species-habitat responses using generalized additive models. *Journal of Applied Ecology* **39**, 755–771.

Sullivan CS (1930) Quis custodiet ipsos custodes? *Emu* **30**, 101.

Tigar BJ & Osborne PE (2000) Invertebrate diet of the houbara bustard *Chlamydotis undulata macqueeni* in Abu Dhabi from calibrated faecal analysis. *Ibis* **142**, 466–475.

Tourenq C, Combreau O, Lawrence M & Launay F (2004) Migration patterns of four Asian Houbara *Chlamydotis macqueenii* wintering in the Cholistan Desert, Punjab, Pakistan. *Bird Conservation International* **14**, 1–10.

Tourenq C, Combreau O, Lawrence M, Pole SB, Spalton A, Xinji G, Baidani MA & Launay F (2005) Alarming houbara bustard population trends in Asia. *Biological Conservation* **121**, 1–8.

Vaill G (2006) Poem. In: *Liberating the Limerick*. (Ed. EW Lefever). Hamilton Books, Lanham, Maryland, USA.

Van Heezik Y & Seddon PJ (1999) Seasonal changes in habitat use by Houbara Bustards *Chlamydotis (undulata) macqueenii* in northern Saudi Arabia. *Ibis* **141**, 208–215.

Walsh GL (1979) Mutilated hands or signal stencils? A consideration of irregular hand stencils from Central Queensland. *Australian Archaeology* **9**, 33–41.

White DM (1985) A report on the captive breeding of Australian bustards at Serendip Wildlife Research Station. *Bustard Studies* **3**, 195–212.

Wills W (1863) *A Successful Exploration Through the Interior of Australia from Melbourne to the Gulf of Carpentaria.* From the journals and letters of William John Wills. (Ed. W. Wills). Richard Bentley, London.

Yang WK, Qiao JF, Combreau O, Gao XY & Zhong WQ (2002) Display-sites selection by houbara bustard (*Chlamydotis [undulata] macqueenii*) in Mori, Xinjiang, People's Republic of China. *Journal of Arid Environments* **51**, 625–631.

Young B, Smith M & Helman M (2005) Captive breeding program: Australian bustard. *The Bird Observer* **834**, 16–18.

Ziembicki M (2007) Australian bustard. In *Lost from our Landscape: Threatened Species of the Northern Territory.* (Eds J Woinarski, CR Pavey, R Kerrigan, I Cowie and S Ward) pp. 184–185. Northern Territory Department of Natural Resources, Environment and the Arts, Darwin.

Ziembicki M (2008) Talking turkey. *Wingspan* **18**, 24–27.

Ziembicki M (2009) Ecology and movements of the Australian Bustard *Ardeotis australis* in a dynamic landscape. PhD thesis. University of Adelaide, Adelaide.

Ziembicki M & Woinarski JCZ (2007) Monitoring continental movement patterns of the Australian Bustard *Ardeotis australis* through community-based surveys and remote sensing. *Pacific Conservation Biology* **13**, 128–142.

INDEX

OTHER TITLES IN THE
AUSTRALIAN NATURAL HISTORY SERIES:

www.ingramcontent.com/pod-product-compliance
Lightning Source LLC
Chambersburg PA
CBHW041130280526
45792CB00013B/2371